ON ART AND THERAPY

ON ART AND THERAPY

An Exploration

MARTINA THOMSON

FREE ASSOCIATION BOOKS / LONDON / NEW YORK

First Published in 1989 by Virago Press

Published in 1997
with a new Foreword and Postscript by
Free Association Books Ltd
57 Warren Street, London W1P 5PA
and 70 Washington Square South,
New York, NY 10012–1091

ISBN 1 85343 366 7 hardback
     1 85343 368 3 paperback

A CIP catalogue record for this book is available from
the British Library.

Produced for Free Association Books Ltd by
Chase Production Services, Chadlington, OX7 3LN
Printed in the EC by J.W. Arrowsmith Ltd, Bristol

CONTENTS

Foreword / vii
Acknowledgements / ix
Introduction / 1
The Teaching of Art / 5
E. M. Lyddiatt / 14
Edward Adamson and Before / 22
Drawing and Painting in Psychotherapy / 29
Picasso and Wölfli / 35
Aesthetic Evaluations / 37
Communication and Response / 40
Being Seen – D. W. Winnicott / 42
Formlessness: An Essential Stage / 46
Training at St Albans / 49
A Brief History of Art Therapy / 54
The Blank Page / 60
The Creative Process – Ehrenzweig / 62
The Encounter with Art Materials / 68
Colour – Within and Without / 74
Modelling and Models / 86
Lines and Boundaries / 89
Recognition / 94
Words and Pictures / 99
Symbol and Image / 104
Sources / 109
Postscript / 118
Postscript sources / 132
Useful Addresses / 133
Index / 134

And wisdom is a butterfly
And not a gloomy bird of prey.
                    W. B. Yeats

 *FOREWORD*

There is a form of human intelligence that is purely visual. We say that the artist works instinctively because we have no way of describing the most subtle of visual choices with words. We recognise the forms but cannot name them. 'I can never remember names but never forget faces.'

Martina Thomson writes about the complex subject of healing by making art, with the experience and insight of a practising artist. She introduces the reader to an adventure that may unlock imagination and transcend even the pain of mental illness.

'Rarely comest thou, spirit of delight'; fulfilment too is rare. Martina Thomson knows of the promise but also the fear of the blank sheet, within which may be found the shapes and colours of conception and death; the before and after of all things. Implicit in her account of her art therapy is the knowledge that some of her patients may look upon the paintings and objects they have made, recognise them and say 'I have made something that is of myself.'

Bernard Cohen
Slade Professor
University College, London

For David

ACKNOWLEDGEMENTS

I wish to thank all those people whose works and writings have inspired me and whose names appear in the Sources at the back. Especial thanks are due to Marion Milner, who took great care over parts of the manuscript I sent her, and to Bani Shorter through whose understanding and encouragement this book has come about.

Warm thanks always to the men, women and children who, over the years, have come to work in the 'art room'.

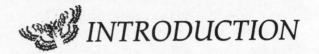 *INTRODUCTION*

*There is no theory that is not
a fragment, carefully prepared,
of some autobiography.*

Paul Valéry[1]

How we come to approach our work is often a matter
of chance. So many things play a part in it: where we
happen to live, where we train and when, how our
background has prepared us, and so on. Chance
encounters thus come to play a part in our thoughts
and reflections and it is for this reason that I must tell
my own story. It will, I hope, be of interest to art
therapists because I have a certain historical perspective. I was initiated into the field by Miss E. M.
Lyddiatt, author of the book *Spontaneous Painting and
Modelling*, one of the pioneers of art therapy. She set
up art therapy departments in many psychiatric
hospitals in this country and my first work was under
her supervision. Her approach was based on Jung's
writings on Active Imagination and she saw her role
as a facilitator and one who 'stayed around'.

Since then I have trained at St Albans and become
acquainted with work in groups and with the often
accepted method of setting a theme or scenario and
of rounding off the session with some verbal explo-

ration of the pictures or models produced. Nowadays there is a widespread tendency to see the art work on a par with every other bit of behaviour enacted during the session as a transference phenomenon and to interpret it as such. We are thus truly on the way towards an art psychotherapy, as is often suggested. Yet it seems to me that in an effort to grasp hold of what it is that is curative in spontaneous creative expression, and produces change in attitude and behaviour, the process that leads to such expression – the painting, the modelling – has been demoted and has become an adjunct to psychotherapy where many therapists feel marginally better equipped to analyse what is going on.

My point then will be to revalue the earlier approach where the involvement with art materials formed the core. I shall consider what it is that made people in psychiatric hospitals turn to painting long before they were encouraged to do so and will investigate what artists have said about what painting means to them. I would like to dwell on what is involved in painting, modelling – in making images.

Many art therapists are artists themselves, and with carrying on their work as therapists they have by necessity demoted their own painting. This has certainly been my experience. What effect does this have on one? And may this in part be responsible – as well as the prescribed training, the growing literature, the recent 'respectability' of the profession – for pushing us into more verbal areas?

AMBIGUITY OF THE TERM 'ART THERAPIST'

When Adrian Hill first coined the term 'art therapy' in 1938, he meant simply that the art does the therapeutic work. He tells us how he spent months in hospital with tuberculosis, how time dragged on and seemed eventually to come to a standstill. One day he took a pencil and made a drawing of the flowers on his bedside table. The drawing somehow intrigued him, he found – 'It had a twist in it which pleased me mightily'[2] – and by the simple act of drawing he had set the pendulum in motion again. As he took more and more to drawing and painting he felt a 'mental emancipation'[3] from his situation. He considered that his drawing and painting helped to pull him through his illness and encouraged by his doctor, he took the good news to the other patients. He urged them to paint also, not to copy but to express in their pictures what they felt about what they were painting, to paint 'dangerously'.[4] Again and again this activity made for an improvement in their condition. Art thus had the power to heal – hence: art therapy.

By the time I first heard of this term it had found a companion: 'art therapist'. This personage complicates matters, for is it now the art or the therapist that initiates the healing process? (It is clumsy when talking about a person to use both the masculine and the feminine pronoun – he and/or she – so I shall use one or the other in turn to stand for both genders.) The ambivalence that is implied in the profession of art therapist – here too we can ask: Is she an artist? Is she a therapist? Is she both? – finds an echo throughout the whole of my life. Perhaps it is not by accident that I have come to adopt a profession that straddles two disciplines.

AMBIGUITY IN MY OWN LIFE

I was born of Austrian parents in Berlin. Our neighbours' daughter, born on the same day as myself and my best friend – we were sometimes dressed as twins – became, at kindergarten age, Aryan while I turned Jewish. But I was not Jewish, I was half Jewish. We left Germany and came to England, so that there were two languages. During the war I was, like many others, an enemy alien and yet a refugee from the enemy. As soon as I began my training as an actress I started attending evening classes in drawing. Summoned to Paris to take part in a television series, I stayed on as a student at the Beaux-Arts. A certain amount of ambiguity is inescapable in life. If I confess to an extreme amount, it is because I wonder whether it makes me particularly sensitive to the issue.

THE TEACHING OF ART

It may be said that art teaching too involves two disciplines, but in my experience that is not so. I was for a time a supply teacher of drawing, for ILEA in adult education. It was life drawing. I saw my task as one of encouraging the student to put down what he saw and valued – what excited, moved him about the object to be drawn. Perhaps more often it was a question of freeing him from what got in the way of this – sometimes a lack of actual looking, sometimes a misjudgement, more often a lack of confidence in what he was doing, which made him labour at an imitation of a 'good drawing' which did not spring from emotional involvement, the 'seed' (Braque[5]) of good work. As it is difficult to keep up a direct involvement for long without let-up – without drifting, doubting, rationalising – I favoured short poses, at least from time to time. They make for urgency and speed up the eye. They make for that vital feeling of being poised and ready.

The focus of my attention seems in retrospect to have moved between the students' drawings and the model. While I came to recognise and value the different ways people drew, I never came to know a great deal about the students themselves.

Recently I have talked to two art students who

both agreed that my approach was roughly what they were getting at their colleges and what was desirable. There was, of course, more technique to be learned in painting – and stamina had to be acquired for long and often boring work – but on the whole, they thought that if you just went on and were encouraged to do so you were bound to improve. I remember too the marvellous hush in the Paris studios where intense work was going on and yet for days on end no tutor came near. In this kind of minimal teaching the tutor can – at least in part – claim the credit for creating an atmosphere conducive to such work.

Both minimal and what I shall call 'maximal' art teachers would see the freeing of the creative ability as their central task. That's a vague enough phrase, but it is often used. On the minimal side we have someone like Ben Nicholson, who believed that only very few lessons were needed on the road to becoming a painter, since painting was really 'the easiest thing in the world' and he regarded every man, woman, child, dog, cat, etc. as a potential artist. 'Teaching art', he said, 'is really a question of discovering the real artistry in a person (everyone has it but often deeply buried) and then liberating this – it is, I suppose, enabling someone (or indeed oneself) to become more fully alive.'[6]

Although 'maximal' art teachers have the same aim, they go about it in a different way. Professor Johannes Itten, famed for his Basic Course at the Bauhaus, bombarded his students with a whole battery of exercises. It was as if he teased out all the experiences and considerations involved in making a picture and presented them separately, one by one, to be consciously pursued. Thus he would start the day with relaxation, movement and concentration exercises and then select a task involving lines, planes, volume, form, rhythm, colour, contrasts or design, etc. – each to be savoured and explored separately. As individual students responded more readily to some of these exercises than to others he began to recognise one predominantly as a 'rhythm' type, another as a light-dark type, and felt he had succeeded 'in opening up individual potentialities'.[7]

Personally I have never been convinced of the value of this kind of programmed learning in regard to art. It strikes me somewhat like a social skills training in regard to life. The learning of vocabulary and grammar should be a by-product of learning a language and surely it does not work the other way round: '. . . true freedom in art is hard-won and art and its freedoms are often confused since the Bauhaus fucked up art education'[8] is how Kitaj put it.

My account of Itten's teaching leaves out the personal factor. He had a devoted following of students and must therefore, I believe, have had an intuitive approach over and above his method. The catalogue of a recent exhibition of his work in his native Switzerland speaks of 'computed intuition'.[9] His book *The Art of Colour* – an extension of the Goethean theory of colour – is widely used in art colleges in this country.

Bomberg was another charismatic teacher although just the opposite of Itten. Like Itten he taught his own practice but by the time he came to teaching at the age of fifty, in 1945, he was himself a mature and powerful painter. His approach was direct: the painter *vis-à-vis* the motif – the landscape, building

or the model. His life-class at the Borough Polytech-
nic in London became famous. The aim was an
almost mystical union with the subject of the painting
or drawing.[10] Each line, each brush stroke, was to
'define the experience of form as well as the form
itself'.[11] The painting was the painting of the discov-
ery of structure, not of structure itself. And more
than structure, what was to be discovered was 'the
spirit in the mass'.

I shall return to that phrase, since in one guise or
another to render 'the spirit in the mass' has always
been an intent of art. For Bomberg it meant 'the
indefinable in the definable' – 'a consciousness of
that which we are all aware of in every manifestation
and on every level of experience'.[12] It was this particular aim he wanted his students to pursue with all
their might. If a student was too tentative, she would
be encouraged to become more involved with the
materials; if the paint became unmanageable, coloured paper was provided to make for clarity; if too
thick a line seemed to imprison the form, experiments with small dots of colour were suggested –
any means of making a mark were permissible. In
contrast to Itten's approach then the means, for
Bomberg, were a by-product of the pursuit of the
artist's purpose.

Frank Auerbach joined the Borough life-class at
seventeen and found the 'atmosphere of research
and of radicalism . . . extremely stimulating. There
was an idea of quality and a lack of fear.' He even
valued the 'anonymous cloudy pieces of paper' that
the people there produced because there was 'absol-
utely nothing cheap or nasty about them'.[13] Yet
Auerbach was rebellious about any teaching and not

inclined to be a disciple. For as John Gross wrote: 'Like all great teachers Bomberg was in danger of enslaving the students whom he had come to liberate.'[14]

I myself had a taste of this at third hand. In the early 1970s I attended a life-class where the tutor was an ex-student of Leon Kossoff, who is an ex-student of Bomberg. Here the drawings were black rather than cloudy grey. The favoured medium was charcoal and this was applied with such vigour that both the pictures and their makers were coated with it. Each stroke was made three times for emphasis: chum, chum, chum – chum, chum, chum – rocking towards and away from the easel. The energy was feverish. A single poised line could have no existence here; willy-nilly you were swept away. It was like taking part in a tribal dance – nor was there a climax or a conclusion. Sometimes the teacher would stand behind one and say 'More so!' And so the pictures were worked over and over.

There was, I believe, an admirable intention behind this drive to keep at it. The pupils were to go far beyond the polite notion of learning to draw. Their intense involvement, possibly their maddening frustration, was to lead them into the realm of art. For my part, when I left the class I felt it had done me some harm – something to do with the trust in my own timing. Is this perhaps the reason why people left the Borough Group and why even Auberbach and Kossoff, who have always acknowledged their debt to Bomberg, did not join the Borough Bottega which superseded it? For 'that is the great thing, to make oneself free of the school and indeed of all schools' (Cézanne[15]).

This 'desertion and rejection'[16] by his students was

a terrible blow for Bomberg. For even while he had doubts about the advisability of teaching art at all, he felt that 'the fundamental creativeness of all true artists, great or small, has been not only in making good works themselves, but in inspiring others to follow them'[17] – to follow the high purposes of art. Teaching at times was his major commitment: he gave his life-blood to it and left off painting for long periods. Often teaching was his only solace, since it gave him a sense of being wanted. When in 1953 his classes were taken away from him and he failed to obtain an Arts Council Retrospective he left for Spain, hoping yet to set up his own school in Ronda. All was ready but few pupils arrived. He and his wife lived in poverty and hardship, but he rallied as an artist. He responded to the grandeur of the landscape in his paintings, painted figures and a last self-portrait; and there were his ultimate compelling landscape drawings. Nevertheless the news from London that he had once again been passed over in regard to a Tate Gallery exhibition had a tragic effect. He spent most of his time writing, going over his life and over and over the injustices he had suffered. He had lost his role as a teacher, there were no students to catch him. He became ill and died neglected in 1957.

POWER IN THE HELPING PROFESSIONS
(Guggenbühl-Craig)

While dwelling on the teaching of art we have come upon two themes which also play a part in art therapy: the theme of 'the spirit in the mass', to which I shall return, and the theme of overdirecting – with its counterside of what this may do to the director.

Formerly the master could claim the work of his apprentices as his own. So for instance in Perugino's workshop the young Raphael was put to work on his master's commissions. But a teacher still, in a way, paints through his pupils. If he can absorb and respond to the different, individual contributions these students make, it will be an enriching experience. If, on the other hand, there is a wish to control the students' development, he must experience anxiety when these students launch out towards unknown goals. In either case a certain loss in creative energy for his own work must be reckoned with.

In his book *Power in the Helping Professions* Adolf Guggenbühl-Craig warns us against trying 'to enforce that which we consider "right" for people'.[18] He says: '. . . the power drive is given freest rein when it can appear under the cloak of objective and moral rectitude.'[19] He also reminds us that we work with our souls, with our selves:

. . . methods, techniques and apparatus are secondary. We, our honesty and genuineness, our personal contact with the unconscious and the irrational – these are our tools. There is great pressure to represent these tools as better than they are . . .[20]

MY INTRODUCTION TO ART THERAPY

While I was teaching drawing here and there in adult education, I was one day sent to take the place of the art teacher in a day centre for the disabled. It was a last-minute replacement and the people whom I was to be with were already assembled and busy. There were six of them, men and women. The routine was to fish out from a pile of old colour magazines a picture that took your fancy and then to make a copy of it in poster colours. The pictures had been chosen by the time I arrived. One was a seascape with a yacht. It impressed me because I ran into trouble when I was asked to lend a hand with the rigging. Everyone was busy painting, yet it was the shared life-experience among the group that mattered most. Then there was satisfaction derived at the end of the day from a picture achieved. 'There you are then,' said one of the painters, very pleased. 'Now when that's dry you put on your varnish and Bob's your uncle, you've got your oil painting.'

I felt that the routine of choosing old, often grubby magazine pictures and the copying business was a bit pathetic, especially because it was a routine. If I had had the chance of working there longer I would have wished to make changes. Yet even so, the choice of subjects, and to a certain extent the way they were copied, touched something of importance in the life of the person concerned. I found myself listening out for this, and reflecting on it with the individual and also the group. Perhaps it was not quite as conscious as it now seems in retrospect, but it was something completely new to me, something that rarely happens in art teaching and something that I liked.

As I talked about my experience at the day centre with friends and acquaintances, I heard for the first time mention of the term 'art therapy'. I heard about the British Association of Art Therapists, which one could join as an associate member and which organised lectures. I joined and went to hear E. M. Lyddiatt, one of its founders, speak at the Royal Free Hospital in Hampstead. I've already referred to E. M. Lyddiatt, who came to play a large part in my life. She was a small, white-haired, quite strong-looking woman with something courageous about her. Her blue eyes were wide open when she looked at you. The pictures that were shown were the first pictures by psychiatric patients I had ever seen.

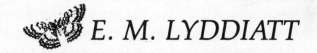 *E. M. LYDDIATT*

Some time after this I wrote to Miss Lyddiatt, telling her about myself and asking how I could get into art therapy. She rang me asking me to come and see her that very day at her house – if I could bear a house in a state of 'upheaval'. As I came to know Lyddiatt's house (hating her Christian name, this is how she liked to be known), I recognised that it was always in a state of upheaval.

It was a house filled with objects, all valued and claiming your attention in all directions: pictures, sculptures, rugs, jugs, plates, ornaments picked up in junk shops, books – predominantly Jungian and anthropological – in piles everywhere and up the stairs, easels, picture frames, patients' paintings and models, and bits of writing – her writing. It was a richly filled house, the coexistence of so many chosen objects making for a kind of order if you allowed half of them to drop into oblivion. That meant that things continually jumped out at you which you hadn't seen before – a bit unnerving, a bit like moving in the realm of the unconscious, a realm Lyddiatt lived close to. 'My house is a mess but my dreams are in order,' she said with a smile. She always went straight for what mattered to her and would nose out quality at once; she liked the direct and primitive and genuine. Sophistication found no resonance in her.

Lyddiatt's house was somehow expressive of her approach to art therapy. She recognised and treasured the value – which to her was synonymous with the therapeutic value – in her patients' paintings and models. She did not at all impose herself as a therapist but allowed the work to happen – and the work that went on under her care was remarkable. She was not inclined to follow the profession to a more sophisticated level, to a more worldly, official and self-conscious stance.

That day when I first came to see her in her house, Lyddiatt asked me to look after her department in a small private psychiatric clinic for some weeks, since for personal reasons she had not managed to get there. All she found time for was her work at the Royal Free nearby where she also went every evening to talk to a young girl going through a psychosis, who would speak to no one else. As Lyddiatt lay dying in the Royal Free five years later, in October 1981, this girl – now a graduate in sculpture – spent long hours day and night at her bedside.

LYDDIATT'S DEPARTMENT

The department I took over, first for a few weeks and eventually for more than seven years, was attached to Bowden House Clinic in Harrow-on-the-Hill, which catered mainly for short-term patients sent there and looked after by their own consultant. About half a dozen consultants worked there. The clinic also had a small geriatric department and altogether there was room for about sixty patients. In 1965 Dr Glyn Davies, the medical director, converted the stables into an art department. Dr E. A. Bennet, the Jungian analyst and friend of Jung, recommended Lyddiatt as the person to set it up and look after it. With her enormous practical energy she furnished the place – two studio rooms, a small kiln room which was also the office and, herself building a low brick wall to restore it, a large and beautiful conservatory. Scouring the junk shops she found heavy wooden tables, both plain and mahogany, old kitchen chairs, stools, ornamental mirrors and jars. She installed a pottery wheel, modelling stands, shelves, as well as geraniums, strange pieces of wood, seashells and stones. There were large supplies of grey sugar paper ('less frightening and more sympathetic than cartridge paper for many people'[21]), red and white clay, some oxides and transparent glazes. Everywhere there were pots with paintbrushes and trays laid out with powder paints and soon many strange models, both drying and fired.

It was a place that provoked dreams, a place where the 'introverted activity'[22] of spontaneous imaginative work was safely housed. The roof leaked, of course, and I must confess it was a bit tatty. This helped to make it a good contrast to the swish and conventional hospital lounge – the only other place where patients could spend time together. People could get the key to work in the art rooms whenever they wanted to. They were also free to come and go on the three mornings a week that the art therapist was present. The work was always individual.

How dared I take on this job? I now think. I suppose it was because Lyddiatt had shown such confidence in me and was always there as a counsellor, a 'supervisor', although at that time I knew nothing about supervision. I could ring her up or go to see her whenever I needed her advice. We went to Bowden House together on my first day and I was amazed to see people she had spoken to in the main building come over and settle down to paint or model as though it was the most natural thing in the world. Sunsets, storms, messes, unconnected squiggles, figures emerged. Lyddiatt took one patient to a room apart for a long talk; others talked more generally or told me about themselves or their pictures. I didn't feel particularly needed or helpful yet it somehow seemed to matter that I was there. When it was time to go people signed and dated their work and put it into their folders or left it out to dry. I was introduced to as many of the staff as possible and then we drove home.

LYDDIATT'S APPROACH

What was the theoretical background of this work as Lyddiatt saw it and as I too understood it at that time? She herself would not have liked to hear it called a theory; to her it was a reality. I shall quote at some length from her book *Spontaneous Painting and Modelling: A Practical Approach in Therapy* – a wise book, now unfortunately out of print – and also from an article she wrote for *The Art Teachers' Journal* which was itself a summary of a paper read at a seminar on 'Art Therapy and the Unconscious'.

I have already said that she found the basis of her work in that of C. G. Jung. Art therapy is 'a waking method of getting in touch with the unconscious so that a new attitude can come into being'.[23] I shall begin as she does by quoting Jung on the unconscious:

We can distinguish a 'personal unconscious', which embraces all the acquisitions of the personal existence -- hence the forgotten, the repressed, the subliminally perceived, thought and felt. But in addition to these personal unconscious contents, there exist other contents which do not originate in personal acquisitions but in the inherited possibility of psychic functioning in general, viz. in the inherited brain structure. These are the mythological associations – those motives and images which can spring anew in every age and clime, without historical tradition or migration. I term these contents the 'collective unconscious'.[24]

Freud too had acknowledged such inherited contents in the unconscious:

Dreams bring to light material which cannot have originated either from the dreamer's adult life or from his forgotten childhood. We are obliged to regard it as part of the archaic heritage which a child brings with him into the world, before any experience of his own, influenced by the experience of his ancestors.[25]

Lyddiatt goes on to quote Marie-Louise von Franz, who saw this 'archaic heritage' as '*the* living creative matrix of all our unconscious and conscious functioning, the essential structural basis of all our psychic life'.[26]

Lyddiatt:

To have the courage, patience and energy to establish a relationship with that living creative matrix – the self we know little about – is the task of art therapy, painting as a means of forging this link being 'an age-old means that is natural to man'.[27]

One seems to pick up a fragment from here and from there, an intimation, a feeling, a thread. Contradictions may seem endless . . . the future all unknown.[28]

It is a deliberate effort to let a mood speak without seeking to control it, and without being overwhelmed by it.[29]

How is this to be done? 'Essentially it is the art of letting things happen',[30] or as Jung says of Active Imagination: 'the task consists solely in observing objectively how a fragment of fantasy develops'.[31] Jung's teachings on Active Imagination grew out of his studies in Word Association, the active element being an engagement with some material – paint, clay, sand, or an activity such as writing, dancing, making music.

In practice a person might start by playing around with a lump of clay. She might push it, squeeze it, and after a while a certain form might suggest itself. She can develop this, destroy it, or allow other forms to take shape. Or a colour might be painted on a piece of paper which may seem to ask for another colour to be intertwined with it, and so on. Sometimes a ready-made image may present itself to be depicted, details of which become clearer or offer suprises as the work proceeds. Lyddiatt encouraged frequent change and trial of different materials – 'perhaps cutting clay that is leather-hard, playing with wood shavings, or graving on plaster'.[32] To her the activity to be stimulated was 'reminiscent of the quiet pondering of adults who in solitude play in streams or wander on the seashore':[33]

Pictures and dreams have a special feeling and this feeling is their value to us. The therapist's job is to help the patient to recognise this feeling, to pick up this quality of the unconscious.[34]

Not only producing the imaginative material but also assimilating it is the crux of the whole technique.[35]

Jung:

. . . the mere execution of the pictures is not enough. Over and above that an intellectual and emotional understanding is needed; they require to be not only rationally integrated with the conscious mind, but morally assimilated.[36]

Yet elsewhere Jung said that the experience itself was the most important thing (i.e. the experience of allowing the fantasy to take shape).[37]

But here we come to the field of creativity in general, where someone like C. R. Rogers can say: 'It is from this spontaneous toying and exploration that there arises the hunch, the creative seeing of life in a new and significant way.'[38] I wish later to go more fully into what is involved in creativity – particularly in the process of painting, since that is what I am most familiar with – and discover how it can possibly be 'curative' in itself.

For the time being I wish to stay with Lyddiatt. She talks of how the image we produce works back on us and affects us 'in the manner of all images'.[39] This is true, of course, and is in fact how the image was produced in the first place, for the to-ing and fro-ing – from painter to picture and picture to painter – is of the essence. Yet in art therapy, probably because pictures are more blindly made, what they say to us when we look at them can be astonishing. Thus a woman I worked with exclaimed: 'I started painting a prison but now I see it's an open door. I feel much better.'

Lyddiatt was well aware of the context in which good work could take place and found active and sustained medical support essential. She did not consider spontaneous painting as a substitute for psychotherapy. She felt that would be an overestimation: 'An ever besetting sin is inflation and it is necessary always to be on one's guard against it'.[40] The 'aim is not to cure or to seek out the reasons for illness, but to set free expectantly the process of active imagining'.[41]

Here is some further advice:

It is important to avoid hurry. One should not demand an explanation. Leave something to the unconscious. It often works in unexpected ways.[42]

Don't expect anything, don't go away.[43]

A sense of humour is invaluable.[44]

I always feel that being with people who are painting is like feeding birds. One moves cautiously lest the 'gift' of painting may vanish.[45]

EDWARD ADAMSON AND BEFORE

From 1950 on Lyddiatt set up departments for spontaneous painting and modelling at the Fulbourn Hospital near Cambridge, at Runwell Hospital, Essex, Leavesden Hospital, Herts, and at London's St Marylebone Hospital for Psychiatry and Child Guidance, at the Halliwick Hospital and the Royal Free. I imagine therefore that in general, in the fifties and sixties, art therapy departments in this country looked somewhat like hers.

Edward Adamson, who has been called 'the father of art therapy' (he was also the founder chairman of the British Association of Art Therapists), opened his studio at Netherne in 1946. It soon became an oasis in the hospital, an 'enabling space' where people might find 'inner resources and give them form in outer reality',[46] a place where patients 'were accorded the dignity of helping to cure themselves'.[47] In his book *Art as Healing* he tells us that he never tries to interpret paintings – he just welcomes them; he sees his role as a facilitator, as a catalyst who allows the healing art to emerge.

SECRET WORK

Adamson discovered that even before he arrived at Netherne patients had taken to drawing – with the charred ends of matchsticks on lavatory paper, with borrowed pencils on the flyleaves of books – and that many of them carried around and treasured little rag-dolls that they had made. Patients naturally turn to making images. We know that, for instance, from the asylum scene in Hogarth's *Rake's Progress*, where the walls are decorated with graffiti and paintings done by the inmates.

Hans Prinzhorn, born in Westphalia in 1886, who studied philosophy and art history in Vienna, singing in London and only later qualified in medicine and psychiatry, completed his collection of works by mental patients in 1921. He was then assistant at the Heidelberg Psychiatric Clinic and acted on the suggestion of its director, Karl Wilmanns, who had started his own collection which he wished Prinzhorn to enlarge and analyse. Prinzhorn thought that 'a primal creative urge' belongs to all human beings and that the sudden activation of this in patients could be explained by the 'inner development or transformation of the patient . . . his autistic concentration on his own person . . . together with his removal from the activities and countless little stimuli of life outside'.[48] Prinzhorn followed up his investigations by a study entitled 'Artistry of Prison Inmates', published in 1926, in which he reproduced photographs of pencil drawings, scratchings and carvings in wood and plaster on cell walls, of small sculptures made out of kneaded bread, playing cards fabricated by the prisoners, and tattoos on their bodies.

The Outsiders Exhibition mounted by the Arts Council at the Hayward Gallery in 1979 showed paintings and sculpture which were selected largely on the criteria used by the French artist Jean Dubuffet for his collection of what he termed Art Brut, Raw Art: 'those forms of extra-cultural creativity which arise from internal promptings, the spontaneous expression of personal resources untainted by outside influences'.[49] Many of the Outsiders exhibited had worked in secret. Gaston Duf, for instance, interned in an asylum in Lille, did drawings on scraps of paper and kept them hidden in his pockets until his doctor discovered them. Henry Darger's thirteen hand-bound volumes of *Realms of the Unreal*, the illustrated adventures of the seven beautiful Vivian girls, were found only after his death. There was something secretive about the whole Outsiders Exhibition. The works were to be shown to the public and then to be withdrawn, or 'eclipsed'. After their 'celebration' at the Hayward a period of 'occultation' was to follow, a period of seclusion in a special 'House' in the tranquillity of the French countryside. The feeling is that there is some honour due to these works, as one might honour a secret, and that they must be actively protected from commercial exploitation. Victor Musgrave – writer, film-maker, director of an art gallery in London – was associated with Jean Dubuffet's venture, and it was at his instigation that the Outsiders Exhibition took place. Musgrave went on to establish an Outsider Archive in this country, a nucleus for a proposed museum of Outsider art.

My point is that such work exists and has always existed before the advent of art therapy, and that Adamson was well aware of this when he took the job of hospital artist at Netherne.

His earlier work (to which I have already referred) had been with Adrian Hill, helping TB patients with painting and drawing. There the aim had also been one of therapy, but Adamson soon came to realise that his work at Netherne was to be radically different. He put this down in part to the far wider range of personalities he now found under his charge. Formerly painting had been regarded as an occupation or a diversion wherein he would assist with technical advice. Confronted by his new patients – who were able to give spontaneous expression to what they found within themselves, and often did so with amazing force – he knew that something quite other was expected of him. He neither criticised nor praised their work: he merely received it.

In my department I found two notices in large writing on sheets of sugar paper. The first was certainly in Lyddiatt's hand; the second might have been written by a patient.
The first reads:

No comments
No questions
No standards

the second:

If you look with the eye of your heart,
you will see on this page a part
of a man's soul.

The respect as well as the kind of protective care this 'art work' calls forth in people who come in contact

with it seems quite general – from Adamson and
Lyddiatt to the organisers of the Outsiders
Exhibition.

DR LEO NAVRATIL AND HIS HOUSE OF
(PSYCHOTIC) ARTISTS

Another attitude to such work is exemplified by Dr
Leo Navratil, medical director of the hospital for
psychiatry and neurology in Klosterneuburg in Austria. Navratil has for years studied and collected the
artistic expressions, particularly the drawings, of the
mentally ill. He follows Freud, who was inclined to
see in the manifestations of psychosis not only the
illness but also an attempt at self-healing.[50] He thus
regards the stereotyped movements of his patients as
a bid to hold on to some certainty, some order, as
well as seeing an ordering principle at work in the
formalism of their drawings. In the spontaneous
creative activities of schizophrenics he sees the same
endeavour as that which underlies all art, namely
'the endeavour to find a meaning in life, to come to
terms with existence, to spin out a new myth which
will help to provide an anchorage'[51] for its creator.
He reminds us that Prinzhorn found 'a metaphysical
bent' to be as characteristic a trait in schizophrenia as
any other. In telling us that one of his own patients
stopped drawing when, after two courses of ECT
treatments, his illness abated, he concludes almost
with regret that 'the artist in him was – the
psychosis'.[52]

In his hospital in Lower Austria he has now set
apart a purpose-built house where his artists live,

work, exhibit and even sell their work. He wants these patients to enjoy in their lifetime some of the honour due to them. I cannot help feeling that Dr Navratil is so enamoured of the strange works of art his patients turn out that he has bedded them down to keep up production. Certainly he holds himself aloof, as does Adamson, from art therapy, from the way it is practised today. But while one knows that Adamson values and relies on the process of painting – 'the conversation that is carried on with the unconscious' – and what this does for the painter, one feels that Navratil is perversely concerned with the product and that he considers the attempts at self-healing involved in paintings, etc. to be phenomena which are by the way and therefore left out of account in the medical treatment of his patients.

While my major task will lie in discovering what it is that is healing or can be harnessed for therapy in the creative process, I must stay a little longer with Dr Navratil. His brief historical survey of schizophrenia included in his *Schizophrenia and Language – Schizophrenia and Art* ends with examples of recent (1966) endeavours in the psychotherapeutic treatment of 'cases' of schizophrenia. He refers to John Rosen in New York, who fed and attended his patients daily for long hours to give them the confidence he felt they had lacked in early life. This allowed him later to voice his interpretations of their hallucinations in terms of suppressed wishes and fears. Navratil also talks of Mme Sechehaye's work with Renée.

Marguerite Sechehaye was a Swiss psychoanalyst who took on Renée when every type of treatment known at the time had been tried, and worked with her for seven and a half years, from 1930 to 1938. She

came to understand that she must satisfy the infantile
needs of this young schizophrenic girl 'symbolically'.
Assuming the role of the good mother, she offered
her apples as 'the good milk from Mummy's
apples',[53] which Renée, leaning against her shoulder,
pressed upon her breast and ate solemnly and with
intense happiness. When Renée's guilt feelings
would not allow her to accept a single expression of
affection Mme Sechehaye provided dolls to whom
they could jointly give the love that was intended for
Renée. Her moving account of this work in *Symbolic
Realisation: a new method of psychotherapy applied to a
case of schizophrenia* has a whole section of Renée's
drawings of her world – diagrammatic line drawings
depicting the trials and tribulations and also the
astonishing insights of 'the little Personage' – which
came to play an important part in the communication
between doctor and patient. At one time Renée
would speak to no one. Her only activity was drawing and writing short notes to M. S. . . .[54] I find it
strange that Dr Navratil, with his specific interest in
drawing, makes no mention of this. But that, I think,
is because while he acknowledges the great achievement of these isolated cases he is as yet doubtful of
the general implications of such dynamically orientated work.

DRAWING AND PAINTING IN PSYCHOTHERAPY

If we are looking at isolated cases that have been successfully treated by imaginative and devoted therapists we might well look at 'Jennifer's' case as reported by the Australian psychiatrist Ainslie Meares in *The Door of Serenity*. In his introduction he tells us that for weeks and months he tried one treatment after another on Jennifer: insulin coma, ECT, narco-analysis, group therapy. Only the last-named seemed slightly to ease her. When the group was disbanded he continued to see her and she continued to be silent, frightened and agitated. One day she fumbled in her bag and gave him 'a weird painting'. From then on it is a story in pictures. *The Door of Serenity* shows us the paintings, each followed by Jennifer's comments and by a few lines of dialogue between doctor and patient, as well as the reflections of Meares at that time and his later reflections when he wrote the book. He learned more and more to understand the symbolism in Jennifer's paintings and his communication with her – if not by silence – was almost entirely in the terms of that symbolism:

On more than one occasion, without looking at me, she offered me a painting covered in red. She would refuse to talk, and I would know from the painting that she was overwhelmed by sexual guilt. On such an occasion my approach to her must needs be extremely remote. I had the warning from other paintings. 'Trees are red.' I had learned that trees are men, and red was sexual.[55]

The pictures, which had been overcrowded and confusing, began to clear. Meares could not say directly: 'You are feeling better.' He had to go via the painting where a bird-form, which had made its appearance early on, was now seen to be flying towards the right-hand side of the picture. Meares had to say: 'The bird is flying towards the right.'

Another 'cure' which proceeded by way of images is described by the analytical psychologist Heinz Westman in *The Springs of Creativity* (1961). He writes:

It was in her drawings, day by day, that the truth of Joan's situation came to light and it was through them that the psyche brought about its own cure.[56]

Her father loomed in the foreground of her existence and it was . . . by means of her drawings that she could begin to see around and behind him.[57]

She . . . considers the feelings of the figures she has drawn and, by doing so, begins a tentative approach to evaluation of her own feelings.[58]

Because Joan's psyche was left free to express itself in its own terms it proceeded to work as the self-regulatory and self-creative agency it truly is and, by means of her drawings, pursued its ontogenetic purpose step by step . . . The psychotherapist was an observer-participant.[59]

Marion Milner came to work with 'Susan' in 1943, soon after completing her psychoanalytic training. It was an analysis that went on for twenty years. In her book *The Hands of the Living God* she describes how after seven years Susan started to do doodle drawings spontaneously in the sessions and how she then took to drawing constantly between sessions too. She brought batches of these pencil doodles into the analysis. Milner was to receive four thousand drawings within the following nine months:

. . . even when the drawings were not interpreted, or not even seen, by me, they did seem to have provided some sort of substitute for the mirror that her mother had never been able to be for her; they did in a primitive way give her back to herself.

[The drawings] had real existence in the outer world and at the same time, in their content and their form, came entirely from herself and her inner world, they were a non-discursive affirmation of her own reality.[60]

I came to see them [the drawings] as my patient's private language which anyone who tried to help her must learn how to read – and speak.[61]

Milner eventually discovered that what was required of her *vis-à-vis* Susan was a certain attitude that lay deeper than words, but I shall come to that later.

Here I would like to dwell once more on the secrecy that surrounds some of the works that people fabricate, which I have touched on earlier in regard to the Outsiders, because there was some secrecy about one of Susan's images too. She had drawn it just after she had had ECT, the night before her first meeting with Milner, but had not mentioned or shown it to her during almost ten years of analysis. Marion Milner was deeply affected when she finally saw it. It showed a figure standing with her head bent down and sideways as if fused with the head of a baby indicated in the cradling arms, head and arms forming a circle. It was poignant, full of anguish, yet also seemed to have in it 'a faint glimmer of hope'. Would Susan one day find in herself the psychic equivalent of the containing arms? It became the central image in the analysis and Milner later thought that the entire analytic work could be seen to revolve around it. Susan's image could be said to have presented the problem and prefigured the solution.

The secret drawing was important in that it contained the seed of Susan's wholeness. In a precarious situation it might have been unwise to expose it. This keeping of your soul, your life, in a safe place is the theme of innumerable folk-tales all over the world, from 'The giant who had no heart in his body' to the beautiful Malay poem of Bidasani who kept her soul in a golden fish. We can read in Frazer's *Golden Bough* how men saw to it that in preparation for a battle or a removal into a new house the soul of the person was hidden away in some safe receptacle, to be reclaimed when the danger was past. Stones, plants or animals which house these souls were honoured, as were sculptures, masks and other cult objects which house spirits in a similar way.

Susan's drawings therefore played an essential part in her analysis – in the process working towards her own self-cure, to use Marion Milner's words. There are of course many more such stories, especially if we include the material in the case histories written by art therapists.

ABOUT WORK PRODUCED IN THERAPY BEING RECOGNISABLE AS SUCH

But it does seem to be true that the work produced in the context of therapy is different from the work that mental patients produce independently. Writing in the 'Outsiders' catalogue about the patients represented in Prinzhorn's collection, Victor Musgrave says: 'Their works were spontaneously generated and did not go through the art therapy process.' And talking about Navratil's artists, he says:

Their work is immediately distinguishable from the vast amount of therapeutic 'art' which, seen in quantity, begins to become somewhat monotonous in theme and content. There is also present a desire to please and satisfy the expectations of the therapist.[62]

Roger Cardinal is well known for his book *Outsider Art*. Writing about the same artists as Musgrave, and in the same catalogue, he says:

Their free artistic self-expression is viewed by Navratil as a valid mode of therapy. It should be noted that Navratil dissociates himself from art therapy as practised in many contemporary hospitals in that he imposes no guidelines, makes a minimum of intervention in the creative process and holds up no standard of achievement to the 'patient'.[63]

'No guidelines – minimum of intervention – no standards' might equally be said to apply to both Adamson's and Lyddiatt's approach – and perhaps to the approach of 'a whole generation of artist-therapists who went into psychiatric hospitals with little if any of the psychotherapeutic preparation

involved in current Art Therapy training'[64] (David
Maclagan in his review of Adamson's *Art as Healing*).
Maclagan, writer, artist and art therapist, considers
that the concern at that time was with the intra-
psychic process rather than with interpsychic or
transferential phenomena, as would be the case with
many art therapists today. Yet as far as the patients
were concerned their painting or modelling in the art
therapy room was 'accompanied'. It was not work
done in secret, or obsessively all day and night, like
Adolf Wölfli's for instance. Nor like Picasso's: 'I paint
the way someone bites his fingernails; for me painting is a bad habit because I don't know nor can I do
anything else . . . it's still often 3 a.m. when I switch
off my light.'[65]

✒ PICASSO AND WÖLFLI

Wölfli was put out by any kind of sympathetic encouragement. He hated being interrupted. And Picasso:

Is there anything more dangerous than sympathetic understanding? Especially as it doesn't exist. It's almost always wrong. You think you aren't alone. And really you're more alone than you were before. Nothing can be done without solitude.[66]

Wölfli often asserted that it was not he who drew his pictures and filled them with writing. 'Do you imagine I could find all that in my own head?[67] he would ask. Similarly, Picasso complained about the connoisseur who was always standing right beside him telling him what to do. ' "I don't like this" or "This is not at all as it should be". He hangs on to the brush . . . certainly he doesn't understand a thing, but he is always there.' It was true to say: 'Je est un autre.'[68]

I have made this comparison between Wölfli, the psychotic artist who spent forty-five years in Berne asylum, and Picasso to show that their work made the same demands on them – they both needed long hours of solitude, they both relinquished sway over their work to 'another', and they both had to protect themselves from sympathetic understanding. It is fashionable to speak of art as a form of communication. Yet while the intention may be one of communication – Wölfli depicted his testament or long reports of fantastic journeys – and writers no doubt want their books published, communication as such does not enter the creative process itself. There the communication is between the painter, say, his subject, his painting, and submerged parts of himself – his unconscious or the 'other' in that sense. I wonder whether it is because the paintings in art therapy are always executed in a communicating relationship with the therapist that they seem to lack a certain core and are distinguishable as 'therapeutic art'. Adamson's collection of works produced at Netherne Hospital, now beautifully housed in an old barn at Ashton in Northamptonshire, abounds with richly expressed individual imagery – often of suffering and suffering resolved. The paintings that stand a little apart are by William Kuralek, who was sent to Netherne because he was an artist but found it impossible to mix with others and had to be given a studio to himself.

What might be inferred from recognisably 'therapeutic' art about the therapy involved in its making? It is an enquiry I shall follow up, but before doing so it may be best to ascertain how such work, as well as the work of mental patients in general, is regarded aesthetically.

AESTHETIC EVALUATIONS

Jung said categorically that the work produced in Active Imagination was not to be regarded as art. That would make for an inflation of the person executing it and set him or her on the wrong track. It will be remembered that Lyddiatt followed Jung's teaching and that her aim was to set free in the people she worked with the process of active imagining. A warning she had written out turned up in my department. It read:

> Leave it flat.
> This is not Art.

I knew what she meant. She was probably trying to save someone from engaging in laborious shading and in calculations of perspective simply for the reason that these were commonly associated with works of art. Perhaps Lyddiatt's message was two-edged in that she felt people would find a more direct expression if they forgot about preconceived ideas. And the more direct and 'primitive' the work was, the more she admired it. Often when we were sorting out pictures she would hold one up and say 'I wish I could paint like that'. Certainly her warning immediately made me think of Matisse's motto: 'Plus c'est *plat, plus c'est de l'art'*, which so often runs through my mind when I am painting.

When Dubuffet started to make his collection of Art Brut at the end of the war in 1945, he went to seek out 'the productions of humble people which have a special quality of personal creation, spontaneity, and liberty with regard to convention and received ideas'. It is interesting that he turned to the art of the mentally ill partly in defiance of the Nazi regime of Occupation with its denouncement and destruction of 'degenerate art'. Over 40 per cent of the Art Brut collection consists of the works of mental patients. Yet Dubuffet insists that a really creative talent is as rare to find in 'insane' persons as in those considered normal. When he goes on to say 'There's no more an art of the insane than there's an art of dyspeptics or of people with sore knees',[69] it is hard to follow him. Psychotic art does recognisably exist. Is it for the reasons Prinzhorn gives – inner transformation of person, autistic concentration on self, removal from life outside – to which I referred earlier? Is it a question of the illness removing certain inhibitions and sharpening sensitivities, in the way that the sense of colour is enhanced in certain altered states of consciousness induced by drugs? Or does psychosis actually activate the creative process – is the creative process symptomatic of schizophrenia, as Navratil believes?

Perhaps one can make a comparison between psychotic art and the art of children, where childhood is the artist as much as the child. And it is certain that the child, like the patient, uses art, as he does play, as a way of exploring and mastering his world. Yet Picasso, who venerated children's art, did not see a comparable exploration in the work of the insane.

The work of a madman is a dead work. It is . . . not an intelligence that progresses and constantly creates in order to progress. The poetry it contains is like the ghost which refuses to give up its corpse.[70]

We, who are so intrigued by the artistry of the mentally ill – and indeed often by their company – are inclined to have lost touch with such a robust attitude. We realise that something about the work may be cramped, one-sided or paper-thin, that it may not unfold or flower, and yet we are moved by it. Indeed, as an art therapist it is difficult not to be overimpressed by abilities that come to the fore.

A severe test of how I myself regard such work faced me when eventually I left Bowden House. There had been a takeover by a new management; the department was to be closed down and I knew that everything would be thrown away. What was I to take with me? From the paintings dating from Lyddiatt's time, sorted through over and over again, I finally left behind all but the works of two painters. I know little about these. One was a young schizophrenic man whose pictures, wherever I came across them – even if sometimes it was no more than a sheet of paper very sensitively covered with one colour – I could not destroy. The other painter was a woman who painted nothing but buildings – towers, blocks, archways – that move me in the way that Morandi's vases and bottles do. Both these painters' work I brought with me, a big pile. In general I found it very much more difficult to part with things done in my time where I saw the pictures or models as a series and recognised the significance of this one and that. It was a terrible wrench leaving any of these behind – and I brought what I could.

COMMUNICATION AND RESPONSE

What then about the therapy involved in works that are recognised as therapeutic 'art'? Obviously it does not follow that because these paintings are not 'Art' but are recognisably paintings done in a therapeutic context they are therefore evidence of a therapeutic process. But I would say that they show evidence of an interpsychic process between painter and therapist, and I believe that people like Adamson are well aware of this. The most striking thing Adamson said to me when I went to see him in his studio in Fulham was that he would never sit down while people were working. He would stand or move about quietly. (Lyddiatt's 'One moves cautiously lest the gift of painting may vanish.') He felt that if he sat down and relaxed the painters would also relax. They would lose their concentration, their thread – like the thread of Ariadne, it seemed to me – whereby their inner vision would find a way out on to the paper. His passivity was active in the sense that he was 'actively vigilant'. Was he the audience to actors on stage, who by his presence and contained response heightened their performance?

What does the actor get from an audience? I think in its participation it reflects his being. My image is of a ray of light bounced back to its source and there making an intense and circumscribed area of light. This helps to keep the actor in his part. Thinking back to my acting career and trying to recall what an audience feels like, I immediately think of scenes where I had to remain silent and motionless, as for instance in the opening scene of Anouilh's *Antigone*. All the characters were ranged on the palace steps while the (one-man) chorus introduced each of us and laid out the plot. I, as Antigone, had just come in from burying my brother Polynices and for a good ten minutes my thoughts dwelt on this in silence. Yet at the same time there was the experience of being in contact with the audience. I recently discussed this with a friend who is an actor and we decided it was most difficult to define. For while one experiences this link with the audience – and can even sense a good, hostile or indifferent one – on no account must one's thoughts be out there; immediately one's performance would be punctured, as it were, and its substance would frizzle away. The audience is an 'observer-participant', which is what Westman called himself in his therapy with Joan.

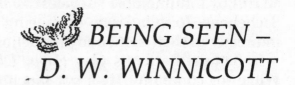

BEING SEEN –
D. W. WINNICOTT

Obviously I am not trying to equate a theatrical performance with a psychotherapeutic session – especially as the actor's fantasies are not awakened in regard to the audience as a patient's would be in regard to the therapist – I simply want to enlarge on what 'being seen' affords. It seems that being seen for what one is – and Stanislavsky calls the ultimate experience in acting the state of 'I am'[71] – being seen in this way has an empowering effect. This is true both in acting and in therapy, and in both it calls forth spontaneity and a trust in unconscious processes.

I am not unaware that there are people who cannot bear being looked at, but the existential psychoanalyst, Ludwig Binswanger, talking about the 'Shame Phenomenon' in *The Case of Ellen West*, points out that 'shame shows to the other precisely what it wants to hide from him, that is, the secret of the existence.' He refers to one of Janet's patients, Nadia. Pierre Janet, a contemporary of Freud, vividly reports how Nadia wished to be invisible and how she kept questioning people compulsively about her looks; she felt misunderstood if people could not see the pimples 'under her skin'. And yet 'She is afraid to be conspicuous to the others, to be different from them, to be less loved by them, and protects herself from all this by innumerable "devices".'[72] Both Meares and Sechehaye found ways of 'seeing' their patients indirectly in their symbolic productions.

In Samuel Beckett's play *Happy Days*, Winnie has Willie for company. He's not much of a companion because he is mostly invisible to her and hardly responsive. Yet in Act II, when she is embedded up to the neck in a mound of scorched grass, she asserts: 'Someone is looking at me still. Caring for me still. That is what I find so wonderful . . . I used to think that I would learn to talk alone. By that I mean to myself, the wilderness. But no. No, no. Ergo you are there.'[73] Beckett's plays have this theme of being seen or heard. Even where the play consists of a monologue – as in *Not I*, where the speaker is reduced to a mouth – it has the attention of an Auditor, a silent figure who is allowed four slight responsive movements. Where there are voices only, as in *That Time*, a Listener is provided. But these people who cannot be seen, who are merely voices or a mouth, speak in the third or second person about themselves, they are unable to say 'I': '. . . for God's sake did you ever say I to yourself in your life come on now (eyes close) could you ever say I to yourself in your life . . .'[74] It seems that people who are not seen may well be denied a subjective feeling of reality.

There is something fundamental about being seen. Much of D. W. Winnicott's writing in the 1960s revolves around a first and fundamental exchange between mother and baby. The baby looks at the mother's face and sees there a reflection of himself, for he sees the mother's face as she reacts to him. Out of this mirroring in the mother and child relationship arises the self-image in the baby. Cherishing feelings in the mother make for warm feelings about himself in the baby. Where the mother is withdrawn the emerging self will feel fragile, unvalued.[75] Winnicott noticed that if the mother's face is unresponsive a baby will approach a mirror with caution.[76] Such 'human' considerations are left out of account in Lacan's observations of the mirror stage.

Jacques Lacan (1901–81), psychoanalyst, thinker, poet, surrealist, brought his subtle insights to bear on the work of Freud, of which his own work was an extension. The mirror phase is his term for that formative event in the infant's life, between six and eighteen months, when it comes to recognise its image in the mirror. By identifying with the image of its counterpart, or with its own reflection in the mirror, the infant apprehends and anticipates, on an imaginary plane, a mastery of its bodily unity. The image here fulfils an organising and constitutive function. By identification with an image a transformation takes place in the subject. Lacan sees this imaginary unification that takes place as the forerunner, or symbolic matrix, of the ego. (Since it is founded on an imaginary basis, the ego here considered would from the start be an ideal-ego [Ideal-Ich] and would not represent the intrinsic subject.[77] Lacan's use of the term 'Imaginary' is highly idiosyncratic. A theoretical development of the theme of the mirror phase, the Imaginary is characterised by the prevalence of the relation to the image by the counterpart [specular ego]. It is one of the three essential orders of the psychoanalytic field: the Real, the Symbolic and the Imaginary.)

That an image, a *Gestalt*, may have formative effects on the organism is borne out in animal ethology; Lacan refers to the fact that the female pigeon relies on seeing another member of its species or its mirror-image for the maturation of its gonad, its reproductive gland.[78] This particular purely visual stimulation is necessary to it for ovulation to take place.[79] More speculatively, the effect of identification with an image on bodily functions is tapped in the experimental treatment of cancer, where the patient's task is to visualise her healthy cells doing battle with and vanquishing the cancerous proliferation. I shall have more to say later about the vital significance of the image.

Winnicott says of the reflection process:

Psychotherapy is not making clever and apt interpretations; by and large it is a long-term giving the patient back what the patient brings. It is a complex derivative of the face that reflects what is there to be seen. I like to think of my work this way, and to think that if I do this well enough the patient will find his or her own self, and will be able to exist and feel real.[80]

FORMLESSNESS: AN ESSENTIAL STAGE

The American psychoanalyst Heinz Kohut, similarly, keeps 'the wish to cure and to help' in check; he aims to reach understanding through empathy – 'Comprehension through the perception of experiential identities', as learnt in primary empathy with the mother.[81] This is going a further step back in the infant's development to the stage of primary narcissism (Freud), Primal Love (Balint) – to the very earliest stage when mother and infant are one. The baby does not perceive the mother, the breast, as a not-me phenomenon: the baby *is* the breast. This diffuse state of being before separation, this merging, lays the foundation for our ability to empathise. It is also the position from which it becomes possible to engage in play and in a creative interchange with the world. Winnicott emphasised its importance. Many patients had the 'need to start from formlessness',[82] 'a non-purposive state . . . a sort of ticking-over of the unintegrated personality',[83] 'a desultory formless functioning – perhaps a form of rudimentary playing',[84] a state where unrelated thoughts need not be organised into a communication.

Masud Khan, writer and psychoanalyst, a collaborator of Winnicott, here has the image of a fallow field. 'Lying fallow' is a mode of being he has come to value in his work; he describes it as an 'alerted quietude', a 'receptive wakeful lambent' state,[85] largely non-verbal and imagistic. His word 'lambent' – licking, as of a flame – moving about as if touching lightly – calls up in my mind the flutter of a butterfly in 'zig-zag wantonness':

> And wisdom is a butterfly
> And not a gloomy bird of prey.
> (W. B. Yeats, *Tom O'Roughley*)

I think Lyddiatt had a feeling for this when she wrote about the quiet pondering of adults playing in streams – when she was content to see her patients playing with wood-shavings. 'Look into the fire, gaze into the clouds, as soon as the presentiments come to you and the voices within you begin to speak, surrender to them . . .' says Pistorius, the teacher in *Demian* by Hermann Hesse.[86] It is known that Leonardo found images by looking at a stained wall. Paul Valéry, examining Leonardo's notebooks, comes to realise 'by what starts and snatches of thought . . . and after what immense moments of lassitude, men are able to see the shadow of their future works . . .'[87] This is so both in art and in science. Since Archimedes, how many good ideas have come to people luxuriating in baths.

In her paper 'The Role of Illusion in Symbol Formation' Marion Milner says that creativity implies 'the ability to tolerate a temporary giving up of the discriminating ego'.[88] The aim is deliberately to allow an illusion, a sense of fusion. The painter puts herself into her subject, is 'fused' with it – i.e. there is a suspension of boundaries between self and not-self. Creativity in general remains closely related to an oceanic undifferentiation. In her work with Susan,

Milner came to understand that she herself was called upon to reach that state: 'a partially undifferentiated and indeterminate state – blankness, an empty circle, emptiness of ideas'.[89] Again and again she had to remind herself not to try and find interpretations but to accept the apparent chaos in order to allow Susan to reach the same state – a state of fusion, of one-ness, from which position alone the work could go forward.

It seems then that both for the artist and for the person in therapy a relationship with this undiffer-entiated state of being is essential. In therapy the dwelling therein needs to be reflected back by the therapist. She must also be trusted to contain the not inconsiderable anxieties which formlessness may generate. The experience 'makes the individual to be, to be found; and eventually enable himself or herself to postulate the existence of the self.'[90]

'SELFING'

As the body, the organism, is continually modified by its interaction with the environment, by an ongo-ing process of exchange between inner and outer, so is the self. When we speak of the self we are referring to a process and not an entity. 'Selfing'[91] would more aptly describe it.

> . . . the pattern is new in every moment and every moment is a new and shocking valuation of all we have been. (T. S. Eliot, *Four Quartets*)

In art therapy we are in a unique position to allow the patterns to emerge.

TRAINING AT ST ALBANS

I took on my work at Bowden House perhaps less in the spirit of Lyddiatt – since I knew little about Jung – and more in the spirit of Adamson, who saw his work in terms of releasing the healing art. It went well. Eventually the troublesome roof was mended, the walls were whitewashed and my time was doubled. When I had been there about four years I was made a full member of the British Association of Art Therapists, in the 'grandmother' clause which allowed established practitioners without training to be regarded as professionals. Training was now compulsory for anyone beginning as an art therapist. I, too, was advised to avail myself of this training and therefore enrolled in the first part-time course at St Albans in September 1980.

If I say that I owe a great deal to my training, it is no mere lip-service. I was made aware of ramifications beyond my comprehension. It opened the door to my reading of Freud, Winnicott, Klein, Victor Turner, the anthropologist, Merleau-Ponty and many others. It taught me about working in groups. It allowed wonderful freedom and generous supplies of materials in the time set aside for 'media'. (I think it is called media in the plural simply because of the lavishness and variety of stuff to work with.)

Although there was no brief to the contrary most
people did not do the kind of painting they had been
involved in before. Perhaps because of the general
emphasis on psychological exploration, the work
done in media was quite searching. I gladly aban-
doned my still lifes in oil and worked with excitement
on large drawings of my father and my dreams on
paper stretched against a wall. And, since it was a
part-time course, it was an introduction to the way
other students worked and to the different settings
they worked in. Altogether I learned so much that I
now feel I was a holy innocent before St Albans; I
had worked in a dream.

Yet the teaching of art therapy as a 'subject' left me
dissatisfied. It was done in group workshops where
we were both students and clients. It was directive
art therapy – a task was laid down to which we had
to address ourselves. Each session was neatly structured with a beginning, middle and end. To start
with, then, the group would sit in a circle together
with the teacher-therapist. A response might be
elicited from each member in turn; for instance: 'Say
in a single word how you are feeling', or instructions
might be given for warm-up rituals like 'Massage
your neighbour's back', or 'With eyes closed explore
your partner's face with your hands'. This beginning
part would come to an end with the setting of a
theme to be explored by making an image: for
example, paint something from your childhood,
paint your ideal house, paint three things you would
take with you on a journey, etc. Unless it was a
question of a group painting we would disperse to
different corners of the room and as the time allowed
was always short, thirty to forty-five minutes, a

pensive yet feverish activity would ensue. This was the middle part. Too soon we would be fished out of it to reassemble in a circle to 'share' our images and to account for them. There were questions and comments from the therapist and sometimes from members of the group until the session was brought to a close. Here a final linking of arms around the circle was sometimes encouraged.

What had happened to art therapy, I asked myself, to have been transformed into this charade? For one thing it seemed to have taken the Encounter movement on board, thus making itself even more complex than before. Forerunners of Encounter groups had appeared in the USA around 1946 and had developed rapidly, California being the most fertile soil. These groups were to provide a setting in which ordinary societal constraints on communication were removed so that people were free to speak out and act out and learn from each other. They were laboratories in human relationships where the emphasis soon came to be on the personal growth and development of the group members, who were to live life vividly to the optimum capacity of the human being, to which end basic personality changes often seemed called for. So intently was this aim pursued that some Encounter groups were regarded as places for 'radical surgery'. They snowballed in the mid 1960s, reaching near epidemic proportions in 1967.[92] I found the Encounter business embarrassing, for reasons which perhaps needed investigating, but in our group there was no room to do this since it was not a therapy group as such. Encounter techniques have always struck me a little like wagging a dog's tail in order to make him feel happy. If I embrace my neighbour, will I therefore feel affection for him? This working from the outside in has never appealed to me.

While the objective of the art therapy workshops was still an exploration of images, it was too deliberate, too conscious an exploration. It was like a searchlight trained here and there; it made for a certain amount of reflection, even for a certain amount of playing around the spotlit areas – like writing an essay and hoping for inspiration. Only the medium, the stuff we worked with – paint, plaster, clay, etc. – with its own yield and resistance, helped to drag one a little out of one's depth. This was valuable, but little of the process found reflection in the 'feedback'. The talk centred around the subject matter and implications of the work produced. It pounced too readily on the material – as if it was urgent to turn straw into gold, as if the purpose was to come away with one or two insights. After a time these workshops felt bitty, very different from the experience of the media sessions, which for me added up to something important.

They were different too from some six non-directive art therapy workshops which we had later. Pamela Gulliver, who led these, was trained at St Albans in 1971 soon after the art therapy course was first established. Although she experimented with 'guided fantasy', *Gestalt* and various directive techniques she still based some of her work on her own student experience. 'Mondays', she told me, 'were given up to the learning of art therapy. There was no programme, we simply spent all day in the studio. Edward Adamson, who was then in charge, would come round and look.' When I asked her whether he remarked on the paintings, she thought that he probably said something apt to each student, but she could not quite remember. I have now learnt that I am not alone in valuing the earlier approach. Even at the present moment a similar procedure to Adamson's Mondays has been re-established in St Albans, the only difference being that now the work is discussed within the group at the end of the three-hour session.

A BRIEF HISTORY OF ART THERAPY

What then had happened between 1970 and 1980 to bring in these structured game-playing sessions? I think it was largely a question of politics. When the training course first opened at Hertfordshire College, St Albans, in 1970, it was for a Certificate in Remedial Art. The word 'therapy' was avoided in the title, for it was feared that to make such claims might 'incur the wrath of psychotherapists and doctors' (P. Gulliver). Until then remedial art had been mostly in the hands of teachers, yet it was felt that teachers did not have the special approach the work demanded. The business of the course was therefore both to separate itself from teaching – this was done to the extent of eschewing membership of the British Association of Art Therapists, which had become affiliated with the NUT – and to define itself within and *vis-à-vis* the medical profession.

Inevitably a lot of psychological and quasi-medical subjects filled the syllabus: child development, mental handicap, delinquency, the brain, perception. There was a therapy group run by a clinical psychologist, and the legal aspects of mental illness were taught by a magistrate. The candidates for the course were selected from a wider field than the arts – for example, nurses and social workers could apply. The image of art therapy as it came to be recognised by the DHSS was being defined in those days, and this had to be in terms that could be communicated to other disciplines. It is not surprising therefore that anything 'woolly' was frowned on.

Psychoanalysis became the frame of reference, but an analytical approach is not so easily come by. For want of it, structures and directions, 'games' which investigate definite areas and are certain to involve people, even if only up to a point, must have seemed desirable. We must remember too that this development coincided in time with the advent of the growth movement and its 'life-enhancing' ethics.

The art therapy course at Goldsmiths' College started as an option within the Institute of Education in 1974 and became a separate course in 1976. Whereas the bias at St Albans had at first been on mental handicap (owing to its connection with the many handicap hospitals in the area), Goldsmiths' course was first linked with child psychology. Perhaps because it was the later course and the art therapy image had already to a certain extent been fashioned, the orientation of Goldsmiths' may have always been slightly more towards the arts. The intake of students was confined to art graduates. The staff now are all art therapists, several of whom hold qualifications in psychotherapy as well. The teaching of art therapy as such goes on in both directive, theme-centred groups and in non-directive group analytic workshops which continue through the year.

In 1984 a further course was established within the Department of Psychiatry at Sheffield University. It has a small intake of students and tries to foster their individual approach to therapy. For this reason students are in placement throughout the year and not only receive clinical supervision but also have regular psychotherapeutically orientated hours of supervision with a member of the staff, who are all practising art therapists. This emphasis on supervision in the course follows the northern tradition established by Winnicott and Balint. The workshops are mostly non-directive.

There are also two training centres in this country for the Rudolf Steiner-based method of 'artistic therapy'. Following Steiner's precepts, the student will study the intrinsic qualities of individual colours and forms and the forces inherent in black and white, and geometric line drawings, and learn about their healing potential. Thus equipped, the artistic therapist will use the various elements of painting, drawing and modelling in exercises, chosen from among a repertoire, for the specific needs of the person in therapy.

Although Adamson was the first course leader at St Albans, he did not stay long. He felt out of tune with what I call the political development, perhaps better described as 'professionalisation'. Lyddiatt, too, was doubtful. I remember once using the word 'interesting' about a painting and she jumped on me for talking 'St Albans language'. Equally unreasonable accusations were levelled at her in the St Albans camp, and still are. It was perhaps an inevitable by-product of professionalisation that the earlier practitioners in art therapy are regarded as woolly.

'PROFESSIONALISM' OUSTED 'WOOLLINESS' AT SOME COST

It is my contention – and I have already laid the ground for this – that a certain woolliness, perhaps more truly a woolly-seeming-ness, may have positive value. 'The art therapist struck me as woolly, vague and edgeless, and the lack of shape, direction and order in her department irritated me. But in looking back I see that in my insanity it was just those qualities that I needed', wrote one of Lyddiatt's ex-patients.[93]

I can see that the setting Lyddiatt created at Bowden House was ideal for the kind of work she believed in, work that would gather itself of its own accord. In other situations – like the day hospital where I am now employed, where different groups come for only two hours or so – it is much more difficult to withhold directions. After all, there is a risk that people will spend the time (and money) doing nothing. I grant that themes given by the art therapist may help to weld groups together, for which there may be an overriding need, especially when they are given spontaneously, intuitively in response to certain preoccupations within the group. Also it is hard to resist the ubiquitous 'What have you got for us today?' by which patients cast you in the part of the good mother. We have seen that there are times when patients need 'feeding' (Sechehaye), but on the whole whenever I have succumbed to such demands I have had reason to regret it – the painter has thereby been robbed of a truly creative experience.

58 On Art and Therapy

I suppose it would be somewhat blind to imagine that a person coming to art therapy does not feel a certain pressure to get busy with the materials. Even the mildest art therapist has at times been experienced as a bully. When Gerry McNeilly, senior art therapist and lecturer in art therapy, tells his groups 'I would like you to use the materials in this room to express whatever you wish',[94] it is still a directive to be carried out. The foreknowledge that the images produced will later be shared and commented on within the group itself imposes a brief. If it were not such a natural, likely thing for people to play with stuff like clay, to make marks and images, there would indeed be no non-directive art therapy.

I feel much can be learnt by art therapists from Winnicott's report of his three-hour session with a certain woman patient in Chapter 4 of *Playing and Reality* (the chapter headed 'Creative activity and the search for the self'). His ability to stay in the present, to accompany whatever is going on in the patient, has to me a feeling of oriental wisdom about it. As the patient moves from chair to floor to window and so on, silent, weeping or throwing out haphazard remarks, he sometimes relieves himself by writing down interpretations that well up in him. But he does not interpose himself.

It reminds me of things I have read of the practice of meditation where the devotee is told 'to go with the pattern . . . neither trying to defeat an enemy or achieve a goal . . . just going along with each situation'.[95] It reminds me too of Soetsu Yanagi's advice in *The Unknown Craftsman* (1940):

First put aside the desire to judge immediately; acquire the habit of just looking. Second, do not treat the object as an object for the intellect. Third, just be ready to receive passively, without interposing yourself. If you can void your mind of all intellectualisation, like a clear mirror that simply reflects, all the better.[96]

And again:

> The guide does not walk ahead of you, but walks with you.[97]

I doubt that it is really necessary, as McNeilly believes, 'to maintain the conductor's baton to guide the group'.[98] The important thing – to stay with the musical analogy – is to have a good ear.

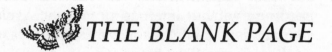

McNeilly makes the tempting suggestion that if a sheet of paper is left blank throughout the session it may be 'the poignant equivalent of silence'.[99] I think it really may be, but would like to fill in what can be contained in such silence. Lyddiatt, for instance, remarks:

It is agreeable to see people occupied and so it may be difficult to accept the fact that some come to a hospital art department for weeks or even months and apparently do nothing. Yet years later these patients have expressed appreciation of coming to paint at the particular time when, as it seemed, they did nothing. Sometimes, in fact, they have become so prolific and infectious in their enthusiasm that it has required effort to remember that it was about these very people one had wondered whether there was any point in their coming.[100]

A writer's block – painful as it is to sit in front of the empty page – may have positive consequences. The writing that is blocked has a chance to 'deepen'. In his re-evaluation of boredom and the writer's block Dr Stephen Simmer of Syracuse University, New York tells us that Bodhidharma, the first patriarch of Zen Buddhism, taught a meditation called 'wall-gazing' and that he is said to have sat before a wall for nine years without interruption. Simmer would recommend the writer to practise a similar vigilance before his block.[101] If the blank sheet of paper at the end of an art therapy session is taken to represent silence, it may represent a most valuable silence.

THE CREATIVE PROCESS –
EHRENZWEIG

But let us then assume that the silence has been broken, that a mark has been made on the paper, an impression in the clay, that a colour has been mixed. This action having been taken, there is now something there which will work back on the person. The first mark will call for a second, and so on. The idea of an image, 'I will paint my mother', may have been a starting point. A vision, something seen, or seen in the mind, may be pursued through its various stages of coming, disappearing and appearing again as the picture is painted. But pictures are not made with ideas but with paint. This is borrowed from Mallarmé's remark to Degas: Degas came to him and said he would write some sonnets; he wasn't short of ideas. 'But Degas,' rejoined Mallarmé ('with his gentle profundity', says Valéry, who tells this story), 'you can't make a poem with ideas. You make it with words.'[102]

With the emphasis on the image, its meaning and interpretation, the process and means by which it is made has been accorded too little attention in art therapy. It is in this very process and in the use of materials that the painter 'comes up against it', that she encounters something other than her conscious will, something which demands consideration, adjustment and a readiness to change course. The deeper the involvement, therefore, the better. It may be necessary to show people the various possibilities of the materials, or at least to encourage them to experiment. One must put over the point that everything is permissible, everything which the picture calls for. Every instinctive prompting in regard to the work should be followed. This is always a response to the work as it progresses and must not be confused with 'Express your feelings!' which elicits a conscious, one-track and often cramped endeavour. If flexibility is a measure of health, it is just this two-way traffic between impulse and consideration (between the primary and secondary process) that is health-giving.

Here I must acknowledge my debt to Anton Ehrenzweig's psychological studies of the artistic imagination, *The Hidden Order of Art*. I would choose it and Winnicott's chapter as most essential reading for art therapists. Ehrenzweig divides the creative process into three phases. The first phase consists of the projection of unconscious parts on to the picture plane (Bonnard spoke of this first phase as a seduction, a falling in love). The second phase is the integration of this material in the picture by means of 'dedifferentiation' – defined as a 'receptive watchfulness',[103] a 'dispersed attention',[104] the use of the primary process as a precision instrument for unconscious scanning,[105] a 'holding before the inner eye a multitude of possible choices'[106] – until a hidden order becomes apparent. (Jackson Pollock: 'I have no fear about making changes, destroying the image, etc. because the painting has a life of its own. I try to

let it come through.'[107]) Finally there is the scrutinis-
ing of the work, seeing it, coming to terms with it –
its reintrojection into the ego.

Towards the end of his book Ehrenzweig compares
the creative process with that of psychoanalysis,
which is similarly structured. We have the projection
of fragmented unconscious material into the receiv-
ing womb of the analyst, the analyst's holding of this
material in his free-floating attention, and its final
reintrojection in ordered form by the patient – 'the
long-term giving the patient back what the patient
brings'.[108] I would add that these stages naturally
overlap and interweave with each other; the final
stage, particularly, may be an extended process.
Ehrenzweig had such a high regard for the value the
creative process has for mental health ('flexible ego')
and for social adaptation (the adaptive relationship
with the picture as the 'other') that his art students
complained that his teaching was psychotherapy in
disguise.

Bearing in mind that we talk of mental illness when
a personality is hedged in by defences, when fixed
attitudes are played out again and again, and that
the therapeutic aim is to dislodge the status quo, is
there not a certain correspondence between this aim
and the aim of much modern art? Could not the
following statement by Mondrian about painting
equally stand as a statement about the therapeutic
task?:

The important task of all art is to destroy the static
equilibrium by establishing a dynamic one . . . the destruction of a particular form and the construction of a rhythm
of mutual relations . . .[109]

Although Michelangelo wrote that poets and painters have licence to dare, to dare do what they choose,[110] it is only with Cézanne's avowed purpose to 'realise' his sensations in front of nature that the attention of the artist has been turned inwards to his own promptings. He spoke of the *sensations confuses* and elsewhere of the 'instincts'. The increasing emphasis on such considerations in painting went hand in hand with the discoveries of psychoanalysis. Odilon Redon (1840–1916), born a year later than Cézanne and sixteen years before Freud, wrote: 'Nothing in art can be done by will alone. Everything is done by docile submission to the coming of the unconscious.'[111]

THE ARTIST MUST LAY HIMSELF OPEN TO THE UNKNOWN

Lucien Freud, a painter whose work shows minute observation, wrote – surprisingly:

The painter must give a completely free run to any feelings or sensations he may have and reject nothing to which he is naturally drawn. It is just this self-indulgence which acts for him as the discipline through which he discards what is inessential to him . . .[112]

There is also Picasso's 'Painting is stronger than I am. It makes me do what it wishes.'[113] Klee saw the artist as existing within the trunk of a tree:

. . . all he does in his appointed place in the tree trunk is to gather what rises from the depths and pass it on. He neither serves nor commands, but only acts as a go-between. His position is humble. He himself is not the beauty of the crown; it has merely passed through him.[114]

Compare this with the author of *The Cloud of Unknowing*. Such was the mystic's attitude:

Let this thing deal with you, and lead you as it will. Let it be active, and you passive. Watch it if you like, but let it alone. Do not interfere with it, as though you would help, for fear you should spoil it all. Be the tree: let it be the carpenter.[115]

The anthropologist and Jungian analyst John Layard clarifies this in a private letter (unpublished) where he talks about 'two wills':

I once had one of the crucial conversations of my life with a Trappist monk . . . In the course of our talk the word 'will' somehow cropped up. So I asked him 'What do you mean by The Will?' He blushed, and then said modestly, 'What *we* (his co-Trappists) mean by The Will is allowing oneself to be drawn.' It then dawned on me that there are two wills, an 'active' (male/external) and a 'passive' (internal) 'female' one. We, in the modern West, think of the will almost entirely as active – cutting down trees, building houses, bridges, mining, making a career or money, and so on. But there is another 'Will' – allowing oneself to be drawn – to 'God' as the priests would say, but as psychologists (mainly Jungian) we would say 'to the Unconscious', which I in my own language would call 'by psychic consciousness' which has *a will of its own* which is often the very opposite of that wielded by everyday 'ego-consciousness' and very often 'has to be obeyed' against what the Western World would usually call 'our Will'.

Schopenhauer speaks of the 'reversal of the will' which happens both in mystic contemplation – 'The work of grace', 'The new birth' – as well as in the work of the artist:

Common to both is a withdrawal from the interests that normally govern . . . the attention we give to things . . . and a temporary relinquishment of our more usual role as . . . purposive agents. For only thus can . . . situations, life itself, be seen in that 'new way' – independently of the patterns and schemes and set responses of ordinary existence.[116]

In his *Aesthetics* Hegel, like Layard, speaks of two wills, but these for him are the will in its natural particularity, and the will in its spiritual universality (which touches the work of art). He finds that the harsh opposition between the necessity of external nature and inner freedom 'has made man into an amphibious animal living in two worlds'.[117] So, too, Proust, towards the end of *Remembrance of Things Past*, speaks of those surface preoccupations and realities which we falsely call life, and from which art will lead us to return 'to the depths, where what has really existed lies, unknown' – in darkness and silence – waiting to be rediscovered.[118]

For those who have lived in houses with cellars Proust's image can be a frightening one. It transports me instantly to the cellar I knew in early childhood – the dark narrow stairs, the strange acrid smell, the glowing fire of the boiler in front of which our dog died of old age, the huge heaps of coal where, as I learned later, manuscripts lay hidden which the Gestapo in their search failed to find.

THE ENCOUNTER WITH ART MATERIALS

It is frightening for patients, and counter to all the control and defences that have been relied upon, to allow passage to the unknown. The intractability of the materials that a person engages with in art therapy is sometimes, in this sense, the first 'unknown' with a will of its own, and this encounter itself can make for the experience of a new attitude. I shall give three examples of this from my own work.

The first concerns a woman of thirty-two who was anorexic:

Sally talked the whole time as she painted her first picture. 'Sand, sand, sand,' she said, dipping her brush in yellow ochre and covering the whole sheet with that colour in broad, horizontal strokes. 'I want to be buried in sand.' She took a long time over this and then suddenly I saw her dipping her brush into blue paint and a small blue pond appeared on the right-hand side of her picture. And now the most amazing thing started. Using blue, green and white alternately she made the water bubble up from that pond. It was like a happening. Again and again it was covered with sand and again and again it bubbled up like an unquenchable spring.

Later I wrote:

There were the flat, horizontal brush strokes which laid the ground and in contrast there was her painting of water, where her brush was at the beck and call of her every intuition. It was as if – as in an initial dream in an analysis – both the problem and its solution became apparent.

Most of the pictures that followed – there were thirteen in all – were painted in black, red and white. 'The red is the pain, the black is despair.' She used white sparingly to portray her own thin existence. She executed these pictures according to a scheme, with refinements and adjustments made here and there. Always she first laid down the ground and imposed the figure later.

This is not the place to go fully into Sally's case, but considering that she had wanted to die because of her father's approaching death, what Ludwig Binswanger says of his anorexic patient Ellen West is apt: 'Her renunciation of life was the most energetic, decisive and self-willed . . .' He talks about 'the inability' of such patients 'to come to terms'; he says

. . . they persist in suffering because things are not the way they would like them to be . . . Freedom consists in the commitment of the *Dasein* to its thrownness as such, non-freedom in denying it autocratically and violating it on the basis of an Extravagant ideal . . . Extravagance, however, is a man's ignoring of the fact that he has not himself laid the ground of his existence, but is a finite being, whose ground is beyond his control.

Sally, in her schematic pictures, always moved on her own ground. (*Dasein*: Existence, Being-in-the-World, where Sartre speaks of consciousness, Heidegger and Binswanger use the term *Dasein*. *Thrownness*: the temporal-existential 'placing', as well as the

psychic determinism (heredity, climate, milieu, etc.)
of the *Dasein*.)[119]

It was only when on three occasions she painted
water that the pictures led a life of their own. They
tossed Sally about. The sea raged and grew calm, the
sky was luminous one moment and black the next,
the landscape in the distance rose and sank. A
seemingly infinite number of combinations were
played out. The spell under which her schematic
pictures were produced was broken. She found the
act of painting water revitalising. What she eventually came to accept – the lesson she had learnt, and
which she illustrated in her last painting – was that
'life is not all black, it is not all red – it is grey'. And
she went to some trouble to show that the grey was
made up out of a random mixture of colours.

*

An extremely good-looking young man, highly
regarded in his firm, where he was a top sales
representative, had made several attempts at suicide.
He had three broken marriages, he had had unsuc-
cessful operations for stomach ulcers, and he felt he
couldn't go on. He came to survey the art department, didn't think much of what he saw, enquired
about materials and let me know that he would work
by himself in the evening. He would do pottery – he
had never done it before – he would make a Grecian
urn.

When I arrived back after the weekend the art
room was transformed to such an extent that I
couldn't at first think what had happened. The floor
and every single surface was covered with clay, every
tool, every brush. An assortment of paraphernalia

had been brought over from the hospital and pressed into service, even bandages. Yet there on the table, placed to be admired, stood a perfect 'Grecian urn' about twenty-two inches in height.

It was a coil pot made to comply with a template he had cut out of cardboard. Where the pot bulged he had made vertical incisions to remove unwanted clay. Later he wetted the shoulder to fix a row of medallions. With a pot so tyrannically made it is not surprising that cracks appeared, that the medallions fell off in the firing. Eventually there were four such pots, since none had achieved perfection. He kept on trying because he wanted me to remember: 'Of all the people who passed through the department, there was one who achieved a perfect piece of pottery.'

While we doctored cracks and other defects and also dealt with the mess that accompanied his work, I often asked him about his childhood, which he always maintained was happy and unremarkable. His doctor, too, knew no better. One night he dreamt he had made a perfect pot but his little son, aged nine, was inside it and calling out to him. He had to break the pot in order to save the little boy. This dream cracked open his 'manic defences'. We now heard things about his childhood which he must have felt shamed him. It seemed to me that with the smashing of the dream pot, something – maybe his imprisoned self – was released. The change in the patient was remarkable and to an extent it was exemplified in his ability to work more organically with his medium and to incorporate accidents and imperfections. Later, and often, when he came to see me we laughed about the extraordinary 'Grecian' episode.

The havoc created in the art room as a result of this patient's work is more commonly expressed in the work itself. It is remarkable how few people, when encouraged to allow free rein to their instincts in regard to their paintings, etc., tear up or destroy what they make. It does happen, but rarely. A patient once tore up her painting but carefully collected all the pieces and handed them to me for safekeeping. Much more common is the 'mess'.

'But why not a "mess"?' I was pleased to read in Ehrenzweig: 'Any creative thinker who ventures into new territory risks chaos and fragmentation . . . What we need today could be a mess.'[120] I have come to regard the mess as a welcome stage in certain people's work. The ability to tolerate it again implies a relinquishing of rigid controls.

In my (limited) work with children the making of a mess has always cropped up. Here it may be more a reaction to controls from outside – family, school – an expression of stored-up anger, a burst of self-assertion. When I worked with chronically consti-pated children at St Thomas' Hospital, the act of messing was appropriate. (I am acquainted with the view that equates all art work with that first gift: a dirty nappy.) Comparing notes with other art thera-pists who had been involved there with these par-ticular cases, I found that 'messing' came up again and again.

'Rita', a silent girl of eight and a half who had no other psychotherapy, made a remarkable recovery within thirteen sessions. One strand of my work with her consisted just of this physical, gestural engagement with the medium – in her case mostly playing with watery paint under the tap and squeezing out paint from plastic bottles.

She painted, washing her paper under the tap and painting more, again and again. She got involved in this game of coming and going. It was fascinating to watch her caught up in this rhythmic activity.

She wetted the paper . . . and carried on hitting down with her red brush to add more flowers . . .

The Jackson Pollock venture (squeezing paint from plastic bottles) ended in a rich thick painting, a mixture of control and accident, she made it lighter and darker in turn and sometimes called out with delight at some beautiful dribble tracery.

COLOUR – WITHIN AND WITHOUT

COLOUR AND SOUND – THE INFLUENCE OF COLOUR

Could the experience of colour – say the mixing of a beautiful pink which so pleased her – act like a balm on Rita? Music therapy relies on 'the influence of sound on man'.[121] May not the experience of colour – and to an extent, could the touching and shaping of form in the case of modelling – have similar effects? It is known that the Greek modes could summon up particular and distinct responses in the listener: the Ionian mode whetted his dull intellect and provoked desire for heavenly things, the Lydian mode soothed the soul oppressed with excessive care, and so on.[122] The relation in which the different elements of the soul – those of reason, of spirit and of appetite – stand to one another was said to be directly affected by means of music; and virtue was the free expression of the beautiful and 'harmonious' soul. Pythagoras, who discovered the numerical ratios which determine the musical intervals, was led to interpret the world in terms of numbers. Numbers seemed to the Pythagoreans to be the first things, and harmonies are therefore of that order. The arrangement of the heavenly bodies depends on intervals regulated by musical harmony. Harmonious numerical arrangement makes the universe a cosmos, i.e. order. The soul, itself a harmony, stood in relation to 'the music of the Spheres', 'the numerical harmonies of the world-soul'. The effect of whole-number ratios on human beings was thought to be universal – everybody was affected in the same way.

In art therapy we are ready to regard the colours a particular person uses as an expression of his feelings. We are much less alerted to the subtle influences that colour (or form) may exert on such a person. The Steiner-trained 'artistic therapist', on the other hand, will consider it central to his or her task. Patients or pupils will be asked to work in certain colours, mould certain forms in accordance with what has been learnt to be their specific restorative influence.

That such influence comes into play the following excerpt from my records will show:

Yesterday a change happened in B. She was painting calmly with little dabs in her own cool colours – light turquoise blue, darker blue and light lemon yellow. She painted a square box containing peace, harmony, order. Then the red paint got spilled at the table where she was working – a bright red spreading pool. After a while B. dipped her brush in it. (She has never used red before, red means anger to her.) She filled her brush again and again, painting with energetic speed and flowing lines that reached out to the edges of her paper and suddenly seemed alive in a new way. It was as though she had woken up out of a deep sleep or enchantment. It was a change she was aware of and others saw.

It was not, of course, a change that was to be permanent, but rather an experience of freedom which she remembered vividly, and was often to refer to, so that it seemed like a foretaste of wholeness.

Yet while there is evidence that colours affect us according to some general rules, colour preferences seem largely to have become a matter of individual taste.

When in early painting colours were prescribed for certain themes by tradition, they were eloquent with symbolic meaning – the gold and sacred background, the heavenly blue of the Madonna's cloak, her red garment of divine love.

By the sixteenth century we have a painter like Veronese, who defended himself against the Inquisition and claimed an absolute right of pictorial licence. He filled his canvases with crowds of accessory figures for the purpose of disposing of light and colour. The American painter Washington Allston (1779–1843) described his experience when looking at Veronese's 'The Marriage at Cana':

> It was the poetry of colour which I felt, procreative in its nature, giving birth to a thousand things which the eye cannot see, and distinct from their cause . . . I understand why so many great colourists . . . addressed themselves not to the senses merely, as some have supposed, but rather through them to that region of the imagination which is supposed to be under the exclusive dominion of music, which by similar excitement they caused to teem with visions that 'lap the soul in Elysium'.[123]

With the Romantic movement and the overthrowing of classical rules in the nineteenth century, the use of colour becomes more and more arbitrary. Van Gogh

– whose early pictures, even, display an arbitrary play of light and shade – was later free to call himself a 'musician in colour'.

[Music] sets the soul in operation. (John Cage)[124]

The fact that music and colour to some extent share this power has often been suggested. That sounds may be apprehended as colours, and vice versa, under the influence of drugs is well known. Under mescalin the sound of the flute is said to give a blueish-green colour. Théophile Gautier wrote about his impressions while smoking hashish: 'my hearing became enormously keen; I heard the noises of colours; green, red, blue, yellow sounds came to me in perfectly distinct waves.'[125]

The sight of sounds and the hearing of colours exists as a phenomenon. In his philosophical enquiry into perception Maurice Merleau-Ponty considers synaesthetic perception to be in fact the rule, the different senses – each modulating the thing perceived – intercommunicating within the unity of the body: 'The senses interact in perception as the two eyes (each with its own monocular image) collaborate in vision . . . My body is the fabric into which all objects are woven.' It is the one body which makes for the one object of perception.[126]

Scriabin has paralleled sounds and colours in a chart – Kandinsky, who refers to this in his book *Concerning the Spiritual in Art*, describes colours by sounds and feelings to which they relate, for example:

Light blue is like a flute, a darker blue a cello, a still darker the marvellous double bass, and the darkest blue of all: an organ; Red . . . glows in itself . . . does not distribute its vigour aimlessly. Light, warm red . . . gives a feeling of strength, vigour, determination, triumph. In music it is the sound of trumpets, strong, harsh and ringing. A cool, light red contains a very distinct bodily or material element . . . The singing notes of a violin exactly express this in music.[127]

Although Kandinsky says that the emotions which colours 'awaken in the soul' are of a much finer texture than can be expressed in words and that his remarks are of necessity provisional and general only, I feel they are also very personal, as any relationship to colour will be. I myself have never thought of colours as notes produced by certain instruments but have been very aware of the heightening, the sharpening up of colours and the corresponding toning down or dulling – this has always felt to me to be a matter of pitch, which the word toning implies, and also a matter of major and minor keys.

For me the colour blue, ultramarine, or rather a mixture of blues from ultramarine towards cobalt and light turquoise, has an infinitely satisfying effect. It is as though those colours make something reverberate within me or accord with it. (Or, one might say, I have a need for the effect blue has on me.) A friend of mine could summon up a certain green – 'Think of the skin of a lemon mixed with washing blue but it is translucent. Some kinds of leaves are like it when you look through them against the sun.' In this green he felt safe from bullies at school and any other intrusion. He came to regard it as a safe refuge.

The inspiration for Winifred Nicholson's painting

was the sight of the rainbow – towards the end of her life she took to carrying one about with her in her pocket in the form of a prism. Her preoccupation with 'the simplicity of the great white light' in which 'all colour lives'[128] puts one in mind of the theology of colour as found in seventeenth- and eighteenth-century Christian mysticism, where the colours of the spectrum are encountered as veils of the divine essence, as the differentiation of the divine light in its descent into the world. The first coloration falls in the sphere of the angels, and we can read that angels are of different colours according to their predominant qualities:

Some are predominant in the *astringent* quality, and those are of a brownish light . . . Some are of the quality of *water*, and those are light, like the holy heaven . . . Some are strongest in the *bitter* quality, and they are like a green precious stone . . . Some are strongest in the quality of *love*, and those are a glance of the heavenly joyfulness, very light and bright; and when the light falls on them, they look like *light blue*, of a pleasant gloss or lustre.[129]

According to this belief it was thought that a man's soul, too, took on the colour of his particular quality 'in variety according to the state of each person . . .'[130]

What about the patient whose pictures are consistently black? Most art therapists will be familiar with people who, at least for periods, do not use any other colour.

A tall, pale girl, always dressed entirely in black, with long straight jet-black hair, came week after week for several months to the art room, took only black paint and methodically covered the entire sheet with it. Sometimes she asked for a PVA medium to seal down her black paint and also to give it a shiny finish. I found it difficult not to wish her to reach out for the seductive colours around her – it seemed wilful and stubborn to shut out so much. Yet gradually I became attuned to the dark and began to see in it. There were distinctions. The pictures were shiny or matt, dense or cloudy; sometimes a small emblem of skull or lightening appeared, once the silver slither of a crescent moon. Two unconnected lines from the *Tao Te Ching* in Arthur Waley's translation kept running through my mind:

I alone am intractable and boorish . . .
I alone am dark . . .

Since than I have read about the Japanese concept of 'the killing of colour'. It was a concept that arose in a culture that was extraordinarily susceptible to the slightest nuances of colour, with names for more than 170 shades – shades arrived at by the semi-transparent layers and linings and underlinings of garments and where even now, in the age of blue jeans, 'pink-plum', 'pink-plum-layer' and 'fragrant-pink-plum-layer'[131] have meaning. With the advent of Taoist thought this seduction of colour began to be regarded as a serious hindrance. It dispersed any apprehension of the innermost essence of things, and thereby of the primordial unity of all phenomena. A trend towards the elimination of colour made itself felt; colour was killed. But not forgotten. Black eventually came to be seen as the ultimate consummation of all colour in so far as it holds a reminiscence of the colours that lie consumed within it. Black therefore implies an internal presence of colour. The senses are withdrawn to an interior point of focus.

Maybe the black pictures of this girl – and possibly also those of other painters, which are too easily felt to be gloomy and sad – could be seen as a powerful symbol of the inner work she was engaged on. When I consider that she was depressed and disorientated, it may have been just this refusal to be distracted that was crucial to her self-cure. Her professional work later was in connection with colour.

Cecil Collins has for years taught his own 'mythology' of colours to students at the Central School of Art and the City Literary Institute, where I attended his class. He ranges his colours on two sides: black and blue as feminine colours on the left; red, yellow and white as masculine colours on the right. These colours may intermarry and produce a child, and so we have a middle column where violet, green and grey stand between their parents. The relationship and interaction of colours in this structure thus become a matter of ritual enactment and story. As an idea it is stimulating and reminiscent of the way the Pythagoreans regarded numbers, which also married, separated, combined into new entities with utter flexibility.[132] I don't know how far this clarifies or muddles a person's direct engagement with colours. Johannes Itten, whom I have talked about as a teacher earlier, found his students in revolt against a colour harmony exercise he gave them – they simply did not find his combinations harmonious. He then asked them each to paint in whatever combinations they found pleasant. The results were entirely individual and have an extraordinary rapport with the photograph of the student which he published beside each example of subjective colour.

COLOUR AND BODY

The colours we find harmonious, or find ourselves in tune with, are then to some extent within us, are our own and find expression in our being. It is a subtle expression; the three writers below hint at it:

Merleau-Ponty in *The Visible and the Invisible*: '. . . it is a question of that Loyos that pronounces itself silently in each sensible thing, in as much as it varies around a certain type of message . . .'[133]

T. S. Eliot: 'I am moved by fancies that are curled/ around these images and cling;/The notion of some infinitely gentle,/Infinitely suffering thing. (*Preludes*)

The painter Paula Modersohn-Becker: 'The gentle vibration of things – that is what I must learn to express.'[134]

In the sphere of sound I have already touched on the belief that everything in nature possesses its own secret and specific sound and internal harmonics. A 'therapy' based on this belief has been evolved where finding, relating to and voicing one's own inner sound is felt to be liberating. That this work involves the body is obvious, for the body is the sound-box. The relationship between colour and body is less obvious.

Yet it has been verified that colour affects the body. Merleau-Ponty, in *Phenomenology of Perception*, quotes experiments with patients with diseases of the cerebellum or the frontal cortex. In these patients – whose muscular tonicity is not, as in the normal person, adjusted to special tasks – the colours red and yellow, generally speaking, induce an arm movement outwards towards the stimulus; blue and green make for a bending back of the arm towards the body – that is: 'red and yellow favour abduction, blue and green adduction'. This bodily reaction is manifested even where the subject is exposed to the colour for too short a time to see it.

Referring to Heinz Werner's investigations into perceptual processes, Merleau-Ponty compares this effect of colour with that of words subliminally perceived. So the word 'hard', for instance, may be said to ready our body before we have time to read it; that is, 'it produces a kind of stiffening of the back and neck, and only in a secondary way does it project itself into the visual or auditory field and assume the appearance of a sign or word.'[135]

Chromotherapy involves the use of colour to stimulate (red), relax (green), or otherwise affect the body. Reciprocally, bodily responses are involved in the choice, mixing and laying on of colour. Painters speak of gut-feelings that come into play. Susan Bach views paintings as though they are bodies. As an analytical psychotherapist she came upon the diagnostic value of spontaneous painting by chance in her consulting room. At the end of the war she conducted a study group of such paintings, probably the first of its kind, under the auspices of St Bernard's Hospital, Southall, London. Her reading of a paper on the findings of this group in 1951 in Zurich led to her association with the Burghölzli mental hospital there and her many years of research as a consultant at the Neurosurgical Clinic of Zurich University. Working with critically ill patients, mostly children, she has perfected a technique of seeing the body in their paintings with such accuracy – almost 'medium-istically', I am told by someone who worked with her – that she is widely consulted as a diagnostician. She seems to trace the pictures from within.

Looking at a coloured drawing of a red-roofed house with a blue bird on top, she asked the consult-ant: 'If you were the house, and its roof your skull, where would the bird's blue tail point to – how would you interpret the situation diagnostically?' The doctor had the medical answer: '. . . it would reach the hypophysis' (the pituitary body of the brain). He sent at once for the case history and discovered that the child who had done the drawing was one on whom he himself had performed an osteoplastic craniotomy three years earlier. The picture had been drawn the day before the operation.[136]

To check her intuitive responses, Susan Bach has devised a grid system whereby she looks at pictures as if they were divided into four quadrants with positive and negative connotations. She also has very definite ideas about colours and makes an important distinction between light and dark shades. She sees dark colours as having a more earthy, healthy feel to them, the lighter colours as showing a decreasing earthiness and decay. Thus light blue is the colour of distance, light yellow indicates a precarious life-situation. Where there is a question of tumours, these are recognisable in the pictures and are in almost all cases portrayed in what came to be known as tumour-red. This red is seen as a reflection of an acute, consuming state of illness as well as, psycho-logically, a 'burning' problem.

Similarly, the chronically constipated children at St

Thomas' Hospital – in my own experience and that of other art therapists – all had a distinct preference for red. Their doctor, too, spoke about the fact that almost all these children loved wearing red; even their mothers seemed to. I noted that for the first weeks 'Rita' would never take off her red cardigan.

> Red . . . glows in itself and does not distribute its vigour aimlessly . . . contains a very distinct bodily or material element. (Kandinsky)

When I consider that Rita's first picture was of a red boat on top but away from the waves, that she later painted a red house on top but above the hill tops, and still later a girl in a hard red skirt but now, at this stage, surrounded and in contact with a flowery meadow, it seems to me that her pink painting, which so delighted her (and which started my enquiry into colour), may well have been experienced as a confirmation of relief.

While I feel that as an art therapist one can never know and understand enough about colours, one should perhaps take to heart John Cage's advice about new listening to new music: 'Not an attempt to understand something that is being said, for, if something were being said, the sounds would be given the shapes of words. Just an attention given to the activity of sounds.'[137] Paintings demand an attention given to the activity of colour.

MODELLING AND MODELS

While the experience of colour may be an extremely subtle one to put into words, the effect of handling clay, for people who actually like the feel of it, is almost always described as soothing, relaxing. The physical business is immediate – touching, smoothing, caressing, squeezing, pulling, pushing, hollowing out or rounding, cutting or digging into it. The mere rolling of a ball of clay between the palms of the hands may give pleasure. The messy stuff of clay is often rejected by patients who cannot bear 'mess', but its very messiness may contribute to the pleasure it gives to those who work with it – a sort of forbidden pleasure. If the aim of psychotherapy is a getting in touch with the instincts, an engagement with clay is often, from the point of view of the art therapist, the most direct way.

It is also a very suggestive medium, more inviting than the blank page. One man came week after week, not in the hospital setting but on the recommendation of his analyst, to work in clay for the purpose of self-exploration. After cutting off a generous portion – he chose black self-hardening clay – he would often barely touch the piece, but look and see in its shape what was to be brought out. Sometimes this was so clear to him that he could work with one hand, indenting here, pulling out there. At other times it was difficult and 'painful', he said. He found that what emerged was always relevant to his present situation – shell, bull, child, face, woman, mother and child, king, villain, trickster, chasuble. The forms were striking in their freedom – I found them beautiful. What was remarkable was that these statues were always achieved within the time he had, within the hour almost to the minute, and that he always used the entire portion of clay he had allowed himself. He valued the pieces, kept them around in his room, brought out one or the other from time to time for particular attention and called them his actors. There he quite consciously referred to what I had told him about Matisse, who kept a cast of jugs, vases and bottles – also chairs and little tables – which were his 'actors'. He had made a relationship with them and would arrange them for his paintings as on a stage; no, the little pewter jug would not play the lead today.

Models seem to have a special value for their maker – more so, in my experience, than paintings. Is it the satisfaction of having left a personal imprint on some bit of material, as it is nice for a child to scratch his name on the school bench? 'Models can be played with . . .' says Lyddiatt. Adamson had found that patients treasured little dolls they had made. Models cannot be stored in folders like paintings but take up space. Are models so valuable because they may be inhabited?

Gisela Pankov, a German psychoanalyst settled in Paris, exploits this aspect in her work with people with distorted or fragmented body-images. In her paper 'Dynamic Structurisation in Schizophrenia' she describes how plasticine models, made by the patient, become the focus of the analytic sessions. She encourages the patient to identify himself with the model in order to enable him to have, perhaps for the first time, a structure to dwell within where desires and feelings can be located – desires and feelings which can be acknowledged to be directed at another body, the analyst's. Only when this work on the body-image has been done, when the patient can feel embodied and the way is paved for a two-person relationship, does Gisela Pankov* begin on the analysis proper.

*In 'Dynamic Structurisation in Schizophrenia' in *Psychotherapy of the Psychoses*, Arthur Burton (ed.), Basic Books, New York, 1961.

# LINES AND BOUNDARIES

Le peintre enroule déroule
plie détord aplatit
casse éparpille effiloche
fronce festonne tortille
tache taraude ravaude
installe accroche répartit
étire boucle débrouille
désigne lance – at s'en va.[138]

The painter rolls unrolls
folds unfolds flattens
breaks scatters frays
gathers festoons tangles
sets up fastens divides
stretches tightens unravels
designs throws – and goes.
(my translation)

This poem by Jean Tardieu is one of twelve 'variations' on twelve ink drawings by Picasso which appeared together in a small book, image and poem confronting each other. A single line in a drawing may do all the poet affirms; for colour and form are not the only elements that concern us. In our work we have to pay attention to the 'activity', the fluctuation, of force, speed, size, balance and so on, of line.

Line as outline has a special significance. Marion
Milner stumbled on this early on in her enquiry *On
Not Being Able to Paint*. The first drawing that somehow pleased her was one where the lines did not
imprison the objects she had drawn but overlapped,
left gaps and generally exemplified a certain freedom
and ambiguity. They were expressive of an experience of merging with her model, the experience and
mobilisation of what Ehrenzweig calls 'dedifferentiation': '. . . the existence of the boundaries (outlines)
or limiting membranes of the self and its objects are
at once perceived and imaginatively ruptured.'[139]
Marion Milner felt that 'to cling to it [the outline] was
therefore surely to protect oneself against the other
world, the world of the imagination.'[140] That this
world was threatening, that it entailed a risk of chaos
and fragmentation, a dwelling in 'uncertainties, mys-
teries, doubts – without any irritable reaching after
fact and reason',[141] I have already discussed. Perhaps
Milner was the first person to discover a bodily
feeling about this risk.

  She had discovered for herself that 'The reality of
the seen is the felt' (G. H. Mead[142]). As she looked at
her jugs and vases, what these stimulated in her and
what she wanted to express in her paintings were
'feelings that come from the sense of touch and
muscular activity rather than the sense of sight'.[143] It
is as if in vision we leap across to what we see –
caress its contours, feel its weight – or are transported
into it and sense in our own bodies how things lean
this way or that, reach up, spread themselves, touch,
take up space. Sensations such as these inform the
painter's gesture, guide the painter's brush. It is then
not only a matter of temporarily setting aside the

'discriminating ego' but with this goes an almost bodily surrender to the object, to the painting. No wonder Picasso spoke of having the 'sensation of leaping into space' every time he undertook a picture. Bonnard's 'being seduced by his motif' is another way of experiencing this surrender, a way more in line with the passive will – allowing oneself to be drawn.

I once saw a salamander take hold of an earthworm to devour it. It gripped it in its jaws somewhere about the middle of its length. But the earthworm was very long and strong and writhed from side to side, tossing the salamander about with it. The salamander would not let go but took great gulps to get more and more of the worm into its mouth and body; but even the swallowed bit of the worm would not keep still. For a time it seemed as though the salamander was entirely at the mercy of this worm which slung him this way and that, until eventually he managed to swallow it. I thought at the time that this was very much my experience of 'being in the grip' of a picture I was painting. I wanted to paint the picture – I had bitten it off, so to speak – but it turned out that the picture, or rather the unknown elements in myself which guided my painting, had the better of me. I found my experience confirmed when I later read: 'The awful thing is that one is one's own Promethean eagle, both the one who devours and the one who's devoured' (Picasso).[144]

It is perhaps just this bodily involvement with painting that is difficult for the schizophrenic artist. 'Oceanic dedifferentiation is felt and feared as death itself', says Ehrenzweig who sees a relation between psychosis and ego-rigidity.[145] While the psychotic is able to give expression to his inner world, to order and shape it in his art, this art confirms his alienation. Therefore, I think, the rigidity in question features in the third stage of the creative process (and it must be borne in mind that this is an attenuated stage throughout the work). I think it is against the reintrojection of the work into the ego that the doors are shut, maybe because it is experienced as a bodily invasion. That is perhaps why Picasso could say that the psychotic does not create in order to progress, and why we feel a lack of unfolding in such work.

Intensification of contours may be a trait of schizophrenic art, as has sometimes been suggested, but I don't think outlines as such necessarily mean an inability to let go. Where outlines are integrated into the picture plane – as, for instance, in Rouault's paintings and watercolours, where outline incorporates shadow – this is not the case. When Matisse draws a single line across the white paper, to him the fields on either side of it take on a different quality, even a slight coloration – so that his line creates two opposed fields of colour.[146] This is not so objectively, but it shows how much his line interacts with the ground. A body outlined in pencil will take on a warm tinge within a cooler surround. It is the interaction of figure and ground on which I have touched earlier and which is so crucial in painting – perhaps the incorporation of light and shadow would be the equivalent in sculpture.

Paul Klee may have been the inventor of the now common exercise in school art classes of 'taking a line for a walk'. 'It goes out for a walk, so to speak, aimlessly for the sake of the walk.'[147] This is often a way – a not too threatening one – of engaging a person in putting marks on paper. It is in fact the doodle. It may be more playful; it may also lead to an image, a configuration. For Matisse it led to what he called his 'revelation in the post office':

In a post office in Picardy I was waiting for a telephone connection. To pass the time I picked up a telegram form which was lying on the table and with my pen traced a woman's head. I was drawing without thinking, my pen moving as it chose, and I was surprised to recognise my mother's face in its subtlest detail . . . At that time I was still an assiduous student of the old school of drawing, seeking to believe in its rules, those useless leavings of the masters who went before us, in a word the dead part of tradition, in which whatever was not observed from nature, whatever sprang from feeling and memory, was despised and called 'fake'. I was struck by what my pen revealed to me and I understood that the mind which composes must retain a kind of virginity with respect to the elements it chooses, and must reject whatever comes to it through reasoning.[148]

'I was surprised to recognise my mother's face . . . I was struck by what my pen revealed to me' . . . There is always an element of recognition in making an image.

# RECOGNITION

For several years I spent my lunch hours with a woman in her late sixties who suffered from senile dementia. 'I thought it was eighteen, nineteen' (she thought she was eighteen or nineteen years old), she told me, 'but now I realise I am nearly at the end of it.' She had been an international chess champion and a professional oboeist. She was often agitated but often quite peaceful, singing to herself in long sustained notes. She liked to come with me to the art room and we understood each other in a rather honey-drunk fashion, her words inventively circumventing nouns which escaped her. She used to sit and model with great concentration. After half an hour or so she would survey the conglomeration in clay and ask: 'Is it alive?' I would echo her question: 'Is it?' She would then look for some tool or a pencil and make two holes – the eyes. At once a smile of recognition transformed her – now it was 'all right', now it was 'alive'.

I was amazed when my little granddaughter, aged eighteen months, showed the same recognition. She was thumping away at a lump of red clay. The dents she had made happened to be in the position of eyes and mouth. To make this clearer I put white clay into the hollows. Instantly and with great delight she recognised a face. She took it up and tucked it under her arm, where she held it tight. With her free hand she threw the rest of the clay back into the bin, every bit, and walked off carrying her clay doll.

A 'presence' comes through – this is how Bridget Riley experiences the final unpredictable transformation that takes place in the creation of a work: 'A mighty pulse skims through the entire picture plane . . .'[149] She thinks of it as an 'event' with which she identifies. The work is, after all, built up 'on polarities such as static and active, fast and slow . . . which find an echo in our psychic being', as well as on 'repetition, contrast, calculated reversal and counterpoint' which 'parallel the basis of our emotional structure'. There should be, she feels, 'something akin to a sense of recognition in the work, so that the spectator experiences at one and the same time something known and something unknown.'[150]

Schopenhauer considers the work of art, or rather the Idea exhibited by the artist, as 'endowed with a mysterious generative power', and compares it in this respect to 'a living organism which brings forth what was not put into it'.[151] I am inclined to regard Bomberg's endeavour to render 'the spirit in the mass' – although so entirely different in its practical approach from Riley's – as an endeavour, like hers, to make contact with a 'presence'.

'There is only the communication between what sees and what is seen', said Picasso more than once, and this was as true, he held, for the artist as for the least cultivated person.[152]

A little boy, aged six, made up the following poem as he walked in his garden:

La petite fraise montre son nez,
La petite pomme montre sa joue,

La pensée montre ses yeux.

Il y a donc des morceaux d'enfant
dans tout le jardin.[153]

The little strawberry shows its nose,
The little apple shows its cheek,

The pansy shows its eyes.

So there are bits of a child
all over the garden.

(my translation)

And what does a painter communicate with in his painting, or a poet in his poetry? Valéry, talking about Mallarmé, his friend and teacher, pointed to a difference between the older man and himself: for Mallarmé the end was in the artefact, while Valéry's own quest centred on the prospection of the mind: 'For him the work; for me, the self'.[154] But the one reflects the other. 'The inner I is inevitably in my painting', said Picasso.[155]

I have already mentioned Susan Bach's researches in the psychosomatic field. Working with severely ill children – children, therefore, whose bodies claimed attention – she saw the condition of the body reflected in their pictures. Speaking of a particular patient, she asks whether we can take his images 'as an expression – profound and unconscious as it may be – of the boy's search for comprehension of what is happening to him'. She continues '. . . his painting shows that "it knows" in him, with an almost awesome precision, the development of his illness . . .'[156]

Again and again, with many different patients, she comes to the conclusion that 'it knows'. She does not define this unknown authority, but I believe the following is apt. She understands the psyche 'as the psyche of an individual soma, as the carrier of man's inborn quest for meaning'.[157] She also says:

This inner reality (encountered in the pictures) is still actively related to by shamans or at festivals within the life-cycle of present-day communities. Is this not, in the last resort, also the goal of the world religions, the creative ground out of which the work of the artist grows?[158]

Matisse, when asked whether he believed in God, would answer: 'Yes, when I work.'[159] 'As in a dream, these opinions have been stirred up within him', Socrates might say, as he does in the *Meno*.

PLATO, FREUD AND THE COLLECTIVE UNCONSCIOUS

Plato set up a psychological experiment to ascertain whether the soul is immortal; was there an innate knowledge in man regardless of experience? A slave boy is confronted with geometric shapes. The boy, who knows nothing of geometry, yet reveals by his answers to questions put by Socrates that he recognises certain immutable truths of space. Knowledge is therefore a kind of reminiscence – a reawakening, a stirring up – and learning a recovery of our own knowledge. And the theory holds not only about our apprehension of relative magnitudes but also that of absolute beauty, goodness, uprightness and holiness (*Phaedo* 75). We read in the *Meno* experiment that where the boy makes mistakes, they are seen as the right sort of mistakes: mistakes he ought to make, mistakes that are natural to the awakening mind. This is surely reflected in Jung's idea of the archetypes in so far as they are 'instinctive trends' or 'inherited possibilities of representation'.

It is surprising, and not generally realised, that in essence Plato's theory is akin to ideas Freud had at times of an 'archaic heritage' which a child brings with him – and also, of course, akin to Wordsworth's 'clouds of glory' (*Ode to Immortality*). It is also close to what Jung termed the collective unconscious and is the opposite of the idea of the *tabula rasa*.

E. M. Lyddiatt saw the forging of links with this 'archaic' layer of the psyche as the most valuable aspect of art therapy. The resultant images will not only be symbolic of personal content but will also contain symbols of mythical motifs and religion, very much as dreams do. Freud realised that an analyst must be at home in 'the history of civilisation, mythology, the psychology of religion and the science of literature'.[160] Without that knowledge the analyst could make nothing of a large amount of his material. Freud recommended that all these subjects should be included in the analyst's training. The principles of alchemy became an increasingly important frame of reference for Jung. Nowadays it is a great disadvantage not to be versed in what is going on in pop music, and pop culture in general, because here is another collective ground rich in reflections of all the aforementioned.

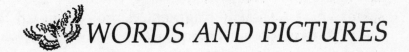# WORDS AND PICTURES

The third part of the creative process – owning up to the work:

Creative man awakens from his oceanic experience to find that the result of his work does not match his initial inspiration. Depressive anxiety is the inevitable consequence. The creative mind must be capable of tolerating imperfections. (Ehrenzweig)[161]

The unconscious material and processes mustered for the making and shaping of the painting, and shown forth in it, are looked at suspiciously in the cold light of day, and even regarded with disgust.

Susan Bach noticed this: '. . . a patient may become unexpectedly disturbed to see afterwards what he has painted, something quite beyond him or perhaps very disagreeable, and may suddenly start to tear or cut it up.'[162] She salvaged the work because of its value for diagnosis and prognosis.

Ehrenzweig is convinced that it is part of the art teacher's task to help his students to overcome their resistance to understanding and accepting their work and not be tempted to sweep it into the dustbin in order to start again with a clean sheet. I have said earlier that he regards this as the third stage intrinsic to the creative process, the stage when the unconscious and necessarily fragmented projections on to the picture plane are taken back into the ego 'in an enriched, more integrated form . . . The external and internal processes of integration are different aspects of the same indivisible process of creativity.'[163] It was one of Bomberg's strengths as a teacher that he was able to alert his students to a recognition and acceptance of what they had discovered or been given; he taught them how to evaluate their work.

This acquainting oneself with one's work is normally a long-drawn-out business. Degas, I believe, kept his pictures for years before parting with them. The whole of the *Picasso's Picassos* exhibition consisted of that significant part of his work which he kept to himself: 'Keeping it physically, closely, near to himself, or locked in places that he knew and owned.'[164] There were studies 'ruminated during long periods, . . . closely connected with the subversive activity of his thought, with problems he had been digesting, playing with, changing and seeing from new angles . . .'[165] Paula Modersohn-Becker writes the following striking passage in one of her letters to her husband, whom she had left in order to paint in Paris. She tells him that she has been 'relatively satisfied' with what she is doing, or 'not quite as satisfied' as before:

This sleeping after each day's work is delightful. My studio is bright in the moonlight. Whenever I woke up, I quickly jumped out of bed to look at my pictures and in the morning my first glance was again directed towards them.[166]

Lyddiatt writes:

When paintings or models have been produced, sometimes – not always – they need to be lived with. To have them around in one's room is a good idea so that one repeatedly comes back to them. Models can be played with and pondered over . . .[167]

Talking about patients' paintings in the last of his Tavistock lectures, Jung recommended that the originals should be given back to the patients:

. . . because they want to look at them; and when they look at them they feel that their unconscious is expressed. The objective form works back on them and they become enchanted. The suggestive influence of the picture reacts on the psychological system of the patients and induces the same effect which they put into the picture. That is the reason for idols, for the magic use of sacred images, of icons. They cast their magic into our system and put us right, providing we put ourselves into them.[168]

Usually nowadays in art therapy work is looked at and discussed at the end of each session. This 'rounding off' is a somewhat forced version of a desirable process, a process to which there may be resistance but which nevertheless, by the very act of painting, has already been set in motion. Ehrenzweig writes: 'The secondary processes of revision articulate previously unconscious components of the work.'[169]

Hegel talks about the need for art where man

. . . draws out of himself and puts before himself what he is . . . he sees his impulses and inclinations . . . outside himself and already begins to be free from them because they confront him as something objective . . . man is released from his imprisonment in a feeling and becomes

conscious of it as something external to him, to which he
must now relate himself in an ideal way.[170]

(For example: the custom of wailing women allows
grief to be contemplated).

Bearing in mind how near at hand these secondary
processes are, they need but the most delicate
acknowledgement on the part of the art therapist,
who may otherwise break trust with the person he
works with. We have seen that the trust built up
between them allows that person to dwell in formlessness, allows her to toy with and follow her
instinctive promptings. If this material is now
pounced on, exploited by the art therapist in order to
prove himself or force insights on to his client, it may
effectively bar the imaginative and emotional process
already stimulated in the client. Masud Khan has a
method of 'uninterpreting'; Winnicott often interpreted in order to reveal the limits of his
understanding.

Not that I think words have nothing to give to
pictures. Matisse is said to have commented that
anyone who wanted to dedicate himself to painting
should first cut out his tongue. It is true that a kind
of dumbness descends over one after long painting;
one just wants to look and look. Yet inner speech
does accompany the making of a work. Picasso, as
we saw earlier, attributed it to the 'connoisseur', the
'other'. It is the conversation carried on with the
unconscious which we value so highly in therapy.
One might even agree that· 'talking and pictures
participate in making each other',[171] and that this
holds good for the creation of a work as well as for
its further life in the public eye.

I feel that words used in the therapeutic session about pictures should stay in tune with the original dialogue between the painter and the painting by which the painting has come about. They may thus echo the elements of surprise and recognition that are inherent in its making. 'I don't know why I put that' is a much-heard and genuine remark:

'I don't know why I put that, but it is part of the picture.'

'Yes, it is part of the picture.'

If words and questions can come out of an empathic feeling for the picture – with awareness of what experiences might have gone into its making – they will best reflect the creative process. This way does not overvalue the picture but values, rather, the life that has brought it into being and which is mirrored therein.

The situation of looking at pictures in a group may or may not involve consideration of the transference, in regard either to the therapist or to different members of the group. Certainly, similarities of theme and treatment which might point to this are often produced unconsciously and may be commented on. The transference interpretation *per se* is widely regarded as 'the only scientific' method of psycho-therapy. My endeavour has been to show what else may underlie recovery and change where art therapy is concerned.

Winnicott's remarks about playing are apt:

It is good to remember that playing is itself a therapy . . . the basis of what we do is the patient's playing, a creative experience taking up space and time, and intensely real for the patient . . . this observation helps us to understand how it is that psycho-therapy of a deep-going kind may be done without interpretive work.[172]

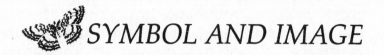SYMBOL AND IMAGE

THE SENSE OF ONESELF AS AN IMAGE

Owning up to pictures is not always a straight-forward business or an easy thing: often screens are produced to avoid it. A middle-aged man, on his second visit to the art room, wrote out with great care a replica of a poster of a 1939 boxing match to which he was taken when he was ten. He said he had decided on this because he didn't want to be 'caught cold' again, as he put it. He had not liked his contribution to a group painting the previous week – a flamelike bullet in red, yellow and black. Even the non-picture 'held more than was put into it'. As he talked about the match – where a crowd of thousands burst through the barrier and invaded the stadium, people who got in free where others had paid as much as thirty guineas for their seats – and as he told us how war broke out only a month or two later, it released then and there (and consequently in other sessions) a flow of images for lack of which the painter had been afraid to be, 'caught cold'. Or, one might say, the images rushed through the barrier of his work.

Images in themselves, once you give yourself to them – once you do not put them by as too ordinary, silly, unimportant – have this property of shifting and of revealing images behind images.

A depressed, middle-aged woman patient playing with some clay produced a model of log upon log upon fence and barricade. At the last minute, and almost as an afterthought, she added a little stick like a small mast. 'Someone is sending a message,' she commented. In her second piece a figure has emerged – herself. Her third model showed her and her family dancing in a circle. Her fourth and last piece was of a maypole dance with a central maypole and streamers and dancers. This last was a collective image which gave her great pleasure. She produced it at a time when she was ready to make a social adaptation.

I used to be put off by the term 'Active Imagination'. I used to press for my fantasy to develop, and nothing happened. Neither is it exactly 'the art of doing nothing'. It is a question of being absorbed and present in whatever is at hand – be it an object, a mental image or the marks of a piece of charcoal on paper.

In his doodle of a woman's head, Matisse was surprised to see his mother's image coming through. To the Lithuanian poet O. V. de Milosz (1877–1939), who wrote in French and who declared he had no country, no home, the word 'mother' conjures up the house of his childhood:

> *Je dis ma Mère. Et c'est à vous*
> *que je pense; O Maison!*
>
> *Maison des beaux étés obscurs de*
> *mon enfance.*
>
> (*Insomnie*)

I say Mother. And my thoughts are
of you, oh House!

House of the lovely dark summers
of my childhood.[173]

Writing about his childhood Seamus Heaney starts:

I would begin with the Greek word *omphalos*, meaning the
navel, and hence the stone that marked the centre of the
world, and repeat it, *omphalos, omphalos, omphalos*, until its
blunt and falling music becomes the music of somebody
pumping water at the pump outside our back door . . .

He goes on to describe the pump – 'a slender, iron
idol' – and remembers how he had watched men
digging to sink the shaft:

. . . that pump marked an original descent into earth,
sand, gravel, water. It centred and staked the imagination,
made its foundation the foundation of the *omphalos* itself.[174]

There is a wonderful to-ing and fro-ing from the
sound of the word *omphalos* to the concrete object,
the pump, which again recalls the *omphalos*, the
navel, the centre. By ruminating on the pump outside
his back door Heaney 'centred and staked' his
imagination; by contemplating their navels the
monks of Mount Athos experienced the divine light
and glory.

Strange how this example of the *omphalos* maps out
the verticality of images – down to the centre, a
sounding of the depths, so to speak. Every picture
tells a story, goes the saying, but stories are strung

along horizontally – 'and then, and then, and then' – like the stories of our lives. The image, like the symbol, contains what it points to; meaning does not follow the image but somehow resonates within it. I have taken poetic images, yet 'the faintest perfume may send us plummeting to the roots of our being'[175] as the taste of madeleine did to Proust. Of such encounters with images we sometimes say: I see it all. It is an experience of initiation, says James Hillman, who calls himself an imaginal psychologist, to make the transition 'from the sense of oneself in a story to the sense of oneself as an image', where all parts are inherent and present at once, where ends and beginnings are held together, 'image-consciousness heals'.[176]

A painting or a drawing of a few bottles by Morandi may be seen as a metaphysical statement. Images have this property of evoking universal resonances. For Nietzsche, 'the self consists of moments in the ring of eternal reoccurrences in which everything is connected'.[177] Maybe this is the healing quality of image-consciousness on which Hillman has come to base his work? It makes our experience into one of *Kairos* – 'a point of time filled with significance – charged with past and future – an instance of temporal integration'.[178]

We may therefore regard the images created in art therapy as soundings (more or less deep) that puncture the experience of everyday time, *chronos* – passing time, waiting time. 'The reverberations bring about a change of being.'[179]

Where Bridget Riley talks about 'an event' (*evenire* – to come out), Hillman speaks of initiation. Both terms imply that something new is recognised and consciously faced. A series of images may then make up a journey.

I would urge the art therapist not to be thrown by her ambiguous presence but in her turn to trust in the creative process and not to interpose herself. She is the guide who walks beside.

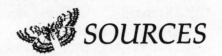 *SOURCES*

1. Paul Valéry, *An Anthology*, selected with an introduction by James R. Lawler, p. 142, Routledge & Kegan Paul, London, 1977.
2. Adrian Hill, *Art versus Illness*, p. 14, Allen & Unwin, London, 1945.
3. ibid., p. 15.
4. ibid., p. 40.
5. *Artists on Art from the 14th Century to the 20th Century*, compiled and edited by Robert Goldwater and Marco Treves, p. 423, John Murray, London, 1976.
6. *Art Monthly*, 72 (1984), p. 6.
7. Johannes Itten, *Design and Form: The Basic Course at the Bauhaus*, rev. edn., p. 8, Thames & Hudson, London, 1975.
8. *Towards Another Picture: an anthology of writings by artists working in Britain 1945–1977*, ed. Andrew Brighton and Lynda Morris, p. 33, Midland Group, Nottingham, 1977.
9. *Die Weltwoche*, 17 September 1971, p. 35.
10. William Lipke, *David Bomberg: A Critical Study of his Life and Work*, p. 97, Evelin, Adams & Mackay, London, 1967.
11. ibid., p. 108
12. ibid., p. 124.
13. *Frank Auerbach*, pp. 20–21, Arts Council of Great Britain, 1978.
14. Lipke, *David Bomberg*, p. 109.
15. *Artists on Art*, p. 367.

16. Lipke, *David Bomberg*, p. 99.
17. Richard Cork, *David Bomberg*, p. 298, Yale University Press, Connecticut, 1987.
18. Adolf Guggenbühl-Craig, *Power in the Helping Professions*, p. 7, Spring Publications, University of Dallas, Irving, Texas.
19. ibid., p. 10.
20. ibid., p. 27.
21. E. M. Lyddiatt, *Spontaneous Painting and Modelling: A Practical Approach in Therapy*, p. 17, St Martin's Press, New York, 1972.
22. ibid., p. 3.
23. *Colour Review: The Art Teachers' Journal*, p. 8, Winsor & Newton, Harrow, Middx, 1976.
24. C. G. Jung, *Psychological Types*, p. 615, Kegan Paul, Trench & Trubner, London, 1923.
25. Sigmund Freud, *An Outline of Psycho-Analysis*, in *The Standard Edition of the Complete Psychological Works of Sigmund Freud*, ed. James Strachey, vol. 23, pp. 166 f., Hogarth, London, 1953–73.
26. Marie-Louis von Franz, *Number and Time*, p.4, Northwestern University Press, 1974.
27. *Colour Review*, pp. 9–10.
28. E. M. Lyddiatt, *Spontaneous Painting and Modelling*, p. 9.
29. ibid., p. 3.
30. *Colour Review*, p. 9.
31. C. G. Jung, *Collected Works*, vol. 13, p. 16, Routledge & Kegan Paul, London, 1967.
32. E. M. Lyddiatt, *Spontaneous Painting and Modelling*, p. 17.
33. ibid., p. 14.
34. *Colour Review*, p. 10.
35. E. M. Lyddiatt, *Spontaneous Painting and Modelling*, p. 9.
36. C. G. Jung, *The Practice of Psychotherapy*, p. 51, Routledge & Kegan Paul, London, 1954.
37. E. M. Lyddiatt, *Spontaneous Painting and Modelling*, p. 10.

38. Carl R. Rogers, *On Becoming a Person*, p. 355, Constable, London, 1967.
39. E. M. Lyddiatt, *Spontaneous Painting and Modelling*, p. 9.
40. ibid., p. 28.
41. ibid., p. 27.
42. ibid., p. 9.
43. ibid., p. 7.
44. ibid., p. 28.
45. ibid., p. 134.
46. Edward Adamson, *Art as Healing* (Foreword by Anthony Stevens), Introduction, Coventure, London, 1984.
47. ibid., p. 2.
48. *Outsiders: An Art without Precedent or Tradition*, pp. 24, 27, Arts Council of Great Britain, 1979.
49. ibid., p. 22.
50. Leo Navratil, *Schizophrenie und Sprache, Schizophrenie und Kunst*, p. 38, Deutscher Taschenbuch Verlag, 1976.
51. ibid., p. 203.
52. ibid., p. 199.
53. M. A. Sechehaye, *Symbolic Realisation: a new method of psychotherapy applied to a case of schizophrenia*, p. 51, International Universities Press, New York, 1951.
54. ibid., p. 46.
55. Ainslie Meares, *The Door of Serenity*, p. 105, Faber & Faber, London, 1958.
56. H. Westman, *The Springs of Creativity*, pp. 255–56, Routledge & Kegan Paul, London, 1961.
57. ibid., p. 193.
58. ibid., p. 226.
59. ibid., p. 256.
60. Marion Milner, *The Hands of the Living God; An Account of a Psychoanalytic Treatment*, pp. 240–42, Hogarth and the Institute of Psycho-Analysis, London, 1969.
61. ibid., p. xxi.
62. *Outsiders*, p. 12.
63. ibid., p. 27.

64. David Maclagan, *Inscape*, December, 1984, p. 15.
65. Dore Ashton, *Picasso on Art: A Selection of Views*, p. 49, Thames & Hudson, London, 1972.
66. ibid., p. 84.
67. Leo Navratil, *Schizophrenie und Sprache*, p. 79.
68. *Picasso on Art*, p. 29.
69. *Aftermath: France 1945–54: New Images of Man*, p. 18, produced for Barbican Centre by Trefoil Books, 1982.
70. *Picasso on Art*, p. 130.
71. Constantine Stanislavsky, *An Actor Prepares*, p. 287, Geoffrey Bles, London, 1945.
72. *Existence: A New Dimension in Psychiatry and Psychology* ed. Rollo May, Ernest Angel and Henri F. Ellenberger, pp. 336–37, Basic Books, New York, 1958.
73. Samuel Beckett, *Happy Days*, p. 37, Faber & Faber, London, 1962.
74. Samuel Beckett, *That Time*, p. 11. Faber & Faber, London, 1976.
75. Christopher Dare and Alex Holder, 'Developmental Aspects of the Interaction between Narcissism, Self-Esteem and Object Relations', *Int. J. Psycho-Anal.*, 62 (1981), p. 327.
76. D. W. Winnicott, *Playing and Reality*, p. 132, Penguin, London, 1974.
77. J. Laplanche and J.-B. Pontalis, *The Language of Psycho-analysis*, p. 251, Hogarth, London, 1980.
78. Jacques Lacan, *Ecrits*, a selection trans. from the French by Alan Sheridan, p. 3, Tavistock Publications, London, 1977.
79. L. Harrison Mathews, 'Visual Stimulation and Ovulation in Pigeons', *Proc. Roy. Soc.* B. 126 (1939), p. 557.
80. D. W. Winnicott, *Playing and Reality*, p. 137.
81. Heinz Kohut, 'Forms and Transformations of Narcissism', *J. of Amer. Psychoanal, Assn.*, vol. 14.
82. D. W. Winnicott, *Playing and Reality*, p. 42.
83. ibid., p. 64.
84. ibid., p. 75.
85. M. Masud R. Khan, *The Hidden Selves: Between Theory*

and Practice in Psychoanalysis, p. 183, Hogarth and the Institute of Psycho-Analysis, London, 1983.
86. Hermann Hesse, *Demian*, p. 103, Grafton, London, 1969.
87. Paul Valéry, *An Anthology*, p. 40.
88. Marion Milner, 'The Role of Illusion in Symbol Formation', in *New Directions in Psycho-Analysis*, ed. M. Klein, P. Heimann and R. Money-Kyrle, p. 97, Maresfield Reprints, London, 1977.
89. Marion Milner, *The Hands of the Living God*, p. 253.
90. D. W. Winnicott, *Playing and Reality*, p. 75.
91. Michael Franz Basch, 'Psychoanalytic Interpretation and Cognitive Transformation', *Int. J. Psycho-Anal.*, 62 (1981), p. 168.
92. Morton A. Lieberman, Irvin D. Yalom and Mathew B. Miles, *Encounter Groups, First Facts*, p. 455, Basic Books, New York, 1973.
93. E. M. Lyddiatt, *Spontaneous Painting and Modelling*, p. 119.
94. Gerry McNeilly, *Inscape*, December 1984, p. 9.
95. Chogyam Trungpa (Rinpoche), *Cutting Through Spiritual Materialism*, pp. 9, 100, Shambala Publications, Boulder, Colorado, 1973.
96. Soetsu Yanagi, *The Unknown Craftsman*, adapted by Bernard Leach, p. 112, Kodansha Int., Japan, 1972.
97. Chogyam Trungpa, *Cutting Through Spiritual Materialism*, p. 21.
98. Gerry McNeilly, *Inscape*, p. 10.
99. ibid., p. 11.
100. E. M. Lyddiatt, *Spontaneous Painting and Modelling*, p. 11.
101. Stephen Simmer, in *Spring 81*, an annual of archetypal psychology and Jungian thought, ed. James Hillman, p. 95, Dallas, Texas.
102. Paul Valéry, *Collected Works*, vol. 12, p. 62, Routledge & Kegan Paul, London, 1960.
103. Anton Ehrenzweig, *The Hidden Order of Art*, p. 105. Weidenfeld & Nicolson, London, 1967.

104. ibid., p. 25.
105. ibid., p. 5.
106. ibid., p. 35.
107. Richard Friedenthal, *Letters of the Great Artists*, p. 272, Thames & Hudson, London, 1963.
108. D. W. Winnicott, *Playing and Reality*, p. 137.
109. *Artists on Art*, p. 428.
110. ibid., p. 69.
111. Klaus Berger, *Odilon Redon*, p. 130, Weidenfeld & Nicolson, London, 1964.
112. *Towards Another Picture*, p. 94.
113. *Picasso on Art*, p. xvii.
114. Paul Klee, *The Thinking Eye: The Notebooks of Paul Klee*, vol. I, p. 82, Lund Humphries, London, 1961.
115. *The Cloud of Unknowing and Other Works*, trans. into modern English with an introduction by Clifton Wolters, p. 101, Penguin, London, 1961.
116. Patrick Gardiner, *Schopenhauer*, pp. 287, 201–2, Penguin, London, 1963.
117. Georg Wilhelm Friedrich Hegel, *Aesthetics: Lectures on Fine Art*, trans. T. M. Knox, pp. 53, 54, Clarendon, Oxford, 1975.
118. Marcel Proust, *A la recherche du temps perdu. Le temps retrouvé*, p. 258, Editions Gallimard, Paris, 1954.
119. Ludwig Binswanger, *Being in the World: Selected Papers*, trans. and with a Critical Introduction to his Existential Psychoanalysis by Jacob Needleman, pp. 252, 258, 321.
120. Anton Ehrenzweig, *The Hidden Order of Art*, p. 147.
121. Juliette Alvin, *Music Therapy*, p. 11, Joseph Baker, London, 1966.
122. ibid., p. 45.
123. *Artists on Art*, pp. 275, 276.
124. *The International Thesaurus of Quotations*, compiled by Rhoda Thomas Tripp, p. 419, Penguin, London, 1976.
125. Gaston Bachelard, *The Poetics of Space*, p. 178, Beacon Press, Boston, Massachusetts, 1969.

126. M. Merleau-Ponty, *Phenomenology of Perception*, trans. Colin Smith, pp. 234–35, Routledge & Kegan Paul/ Humanities Press, New Jersey, 1962.
127. Wassily Kandinsky, *Concerning the Spiritual in Art*, pp. 59, 61–63, George Wittenborn, New York, 1947.
128. Winifred Nicholson, *Unknown Colour: Paintings, Letters, Writings*, p. 101, Faber & Faber, London, 1987.
129. *Color Symbolism*, Six Excerpts from the Eranos Yearbook 1972, Spring Publications, Dallas, Texas, 1977: Jacob Bohme as quoted by Ernst Benz in 'Color in Christian Visionary Experience', p. 106.
130. ibid., p. 111.
131. *Color Symbolism*: Toshihiko Izutsu in 'The Elimination of Colour in Far Eastern Art and Philosophy', p. 170.
132. Anton Ehrenzweig, *The Hidden Order of Art*, p. 288.
133. M. Merleau-Ponty, *The Visible and the Invisible*, ed. Claude Lefort, trans. Alphonso Lingis, p. 208, Northwestern University Press, Evanston, Illinois, 1968.
134. Paula Modersohn-Becker, *Briefe und Tagebuchblatter*, p. 195, Kurt Wolff Verlag, Munich, 1922.
135. M. Merleau-Ponty, *Phenomenology of Perception*, pp. 209, 235.
136. Susan Bach, 'Spontaneous Paintings of Severely Ill Patients', Documenta Geigy, *Acta psychomatica*, p. 23, J. R. Geigy, S. A., Basel, 1969.
137. *The International Thesaurus of Quotations*, p. 419.
138. Pablo Picasso and Jean Tardieu, *L'espace et la flute*, p. 14, *Die Arche*, Zurich, 1959.
139. Peter Fuller, *Art and Psychoanalysis*, p. 171, Writers & Readers, London, 1980.
140. Marion Milner, *On Not Being Able to Paint*, p. 17, Heinemann, London, 1971.
141. *The Letters of John Keats*, ed. Maurice Buxton Forman, p. 72, Oxford University Press, 1935.
142. George Herbert Mead, *The Philosophy of the Act*, p. 10, University of Chicago Press, Chicago and London, 1938.
143. Marion Milner, *On Not Being Able to Paint*, p. 10.

144. *Picasso on Art*, p. 48.
145. Anton Ehrenzweig, *The Hidden Order of Art*, p. 122.
146. Aragon, *Henri Matisse: A Novel*, trans. Jean Stewart, vol. 1, p. 138, Collins, London, 1972.
147. Paul Klee, *The Thinking Eye*, vol. I, p. 105.
148. Aragon, *Henri Matisse*, vol. 2, p. 46.
149. Anton Ehrenzweig, *The Hidden Order of Art*, p. 85.
150. *Towards Another Picture*, p. 120.
151. Patrick Gardiner, *Schopenhauer*, p. 213.
152. *Picasso on Art*, p. xxiii.
153. Etienne Chevalley, *Miracles de l'enfance*, p. 39, La Guilde du Livre, Lausanne, 1952.
154. Paul Valéry, *An Anthology*, p. xviii.
155. *Picasso on Art*, p. 47.
156. Susan Bach, 'Spontaneous Paintings . . .', p. 30.
157. ibid., p. 14.
158. ibid., p. 62–63.
159. Aragon, *Henri Matisse*, vol. 2, p. 221.
160. Sigmund Freud, *Two Short Accounts of Psychoanalysis*, p. 165, Penguin, 1962.
161. Anton Ehrenzweig, *The Hidden Order of Art*, p. 193.
162. Susan Bach, 'Spontaneous Paintings . . .', p. 20.
163. Anton Ehrenzweig, *The Hidden Order of Art*, pp. 104–5.
164. *Picasso's Picassos*, an exhibition from the Musée Picasso, p. 31, Arts Council of Great Britain, 1981.
165. ibid., p. 90.
166. Paula Modersohn-Becker, *Briefe und Tagebuchblätter*, p. 228.
167. E. M. Lyddiatt, *Spontaneous Painting and Modelling*, p. 8.
168. Jung, *Collected Works*, vol. 18, p. 181, Routledge & Kegan Paul, London, 1977.
169. Anton Ehrenzweig, *The Hidden Order of Art*, p. 104.
170. Hegel, *Aesthetics*, pp. 31, 48–49.
171. Nelson Goodman, *Languages of Art: An Approach to a Theory of Symbols*, Harvester, Brighton, 1981, p. 88.
172. D. W. Winnicott, *Playing and Reality*, pp. 58–59.

173. O. V. de L. Milosz, *Oeuvres Complètes*, Vol. 2, *Poésies II*, Editions André Silvaire, Paris, 1960.
174. Seamus Heaney, *Preoccupations*, pp. 17–20, Faber & Faber, London, 1984.
175. Jill Purce, *The Mystic Spiral: Journey of the Soul*, p. 7, Thames & Hudson, London, 1974.
176. James Hillman, *Puer Papers*, p. 122, Spring Publications, University of Dallas, Texas, 1979.
177. Eugen Bar, *Semiotic Approaches to Psychotherapy*, vol. 1 (*Studies in Semiotics*), p. 61, Indiana State University, Bloomington, 1975.
178. Frank Kermode, *The Sense of an Ending*, p. 47, Oxford University Press, 1967.
179. Gaston Bachelard, *The Poetics of Space*, p. 18.

POSTSCRIPT

We must travel abreast with
nature if we want to know her,
but where shall be obtained the Horse –
Emily Dickinson[1]

When deciding to reissue my book almost ten years after it was written, my new publishers felt that it needed a postscript. There have after all been developments in art therapy. Not only has the profession achieved state registration, but it has a code of ethics, a career structure, a place among the clinical professions within medical practice and in the universities. Much has been published on theory and method, both in books and in articles in *Inscape,* the journal of the British Association of Art Therapists, where work in an ever-widening field is described: work in prisons, in schools, in the community, work with people with Alzheimer's disease, with psychotic patients, people with Aids, with learning difficulties, and work with the dying. Many art therapists now go in for an extra training in psychotherapy, some become analysts, and have by this conjunction definitely enriched the field. Art psychotherapy and analytical art psychotherapy are now established; considerations of transference and counter-transference are central. Words therefore have assumed a major role and the art therapist has gained a voice.

At a recent launch of a new book on the subject, I was struck to hear one of the authors say that we, as a profession, want to be heard, our expertise is under-estimated, we do not want to be at the bottom of the pile. And how often have I read that we, as art therapists, do not want to be in the role of handmaidens. This determination assumes an urgency in days of diminished public funding, and confronted as we are with NHS managers who will try eventually to test every therapeutic activity by the yardstick of something called evidence-based medicine, or EBM. A systematic review of experiences gained in therapy sessions and interpretations thereof, a willingness to re-evaluate one's work in regard to such findings, and findings in the field in general, are excellent objectives. The evidence required by EBM, however, is not to be derived from clinical practice but from specifically implemented research involving large numbers (with control groups in place) and must be replicable. Can we rally to that in a discipline where uncertainty is a given, where spontaneity is of the essence, where one is dealing with the mind, the soul? Perhaps the soul were best phased out.

As I am writing this in the run-up to the General Election I hear the word 'evidence' used by all political parties. 'Yes, of course, we will implement such and such, but we have not yet got the evidence.' Evidence will be weighed against evidence, and judgements will be made not necessarily by experts, but by administrators and with a view to cost effectiveness – evidence used as a kind of unwieldy smokescreen. New evidence always comes to light. To survive in the current competitive situation, the profession is called upon to define, evaluate and structure the 'services' it supplies ever more efficiently.

No wonder then that from the current perspective, critiques are levelled at what I wrote so long ago as all very well in its time but now superseded. Yet I feel that as a counterweight to the verbal, as an art-based book,

it may still have something to offer. For, no doubt, the
image and its silent making, though somewhat margin-
alised, remain part of the process. To apprehend both in
their puzzling diversity and to be open to all the
imaginative play they may give rise to, to be sensitive
to the feel and style of the image, must surely most
importantly underlie what goes on in art therapy. I
want therefore to return once more to that familiar
ground, perhaps adding some things which I have
discovered since my book was first published, and want
also to survey some of the work that is going on now,
and touch on what has been written.

The business of eye, hand; hand, eye – reflection and
impulse – is itself revitalising. 'By the hand it unceasingly changes the eye unceasingly changed. Back and
forth ... Truce for a space and the marks of what it is to
be and be in the face of.'

Truce, Samuel Beckett's word,[2] implies a temporary
suspension of conflict and a reciprocal acknowledgement. The outcome of the engagement is not known
beforehand. The marks as they appear on the canvas
draw the painter in, they act and interact with each
other and soon he is busy on their behalf. Unceasing
shifts and changes, a state of flux. With a sort of
receptive attention he is prepared to manoeuvre here
and there, pull this forward, push this back until gradually the thing becomes clearer. It is as if in the course of
painting – a motif, a subject, even a vision – another
subject appeared and that it is the actual subject,
namely the internal relationship of the work.[3] You are
awaiting and attending it – nurturing it as a mother her

unborn child – humbly and patiently; battling at times to set it free, not knowing yet what it is.

'I must impress upon myself that I know nothing at all', writes Degas,[4] 'for it is the only way to progress'. Lately I've been occupied with this thought of the knowledge of knowing nothing, and now have annexed it for art therapy. And it's been a relief to me and I've understood something better than I did before, namely a fear I've sometimes had before a session with patients, the fear of being found out, of being a fraud even. I've heard this feeling echoed by other art therapists who are inclined to blame their particular training, their particular situation. But the reason for it is that all the training, all the traditions and achievements, all the theories, may give one a context for one's work but cannot bolster one up – one stands there empty-handed. One has no horse to ride on. Quite a terrifying experience but perhaps necessary in order for one's responses to be uncluttered, spontaneous, and one's awareness and tolerance of the various things going on to be as wide as possible. Because we cannot know beforehand what a patient will bring today and what will happen between us.

An actor knows the play, has learned and rehearsed his part and yet has stage fright because he knows that all his preparation will not of itself get him there. In order for his performance to *live* he must allow himself to move on the threshold of the unconscious, to be buoyed up by the unconscious. Stanislavsky in his marvellous *An Actor Prepares*[5] writes:

> I want you to feel right from the start, if only for short periods, that blissful sensation which actors have when their creative faculties are functioning truly, and subconsciously. Moreover, this is something you must learn through your own emotions

and not in any theoretical way. You will learn to love this state
and constantly strive to achieve it ... Our freedom on this side
of the threshold is limited by reason and convention; beyond it,
our freedom is bold, wilful, active and always moving forward.
Over there the creative process differs each time it is repeated.

The reliance on the unconscious layer of the mind
was acknowledged, somewhat to my surprise, even in
the Age of Reason, the Enlightenment, which by popu-
lar assumption is associated with a quasi-'religious'
belief in natural science, and in the rational, cognitive
function of the mind and to have ruled out, or understood as malfunction, its irrational workings. For see
what that lovely man Diderot,[6] the Encyclopaedist, says
about his mental processes which he had set himself to
observe. He talks of 'conjectures', of 'sniffing out' new
procedures, of 'divinations', 'extravagances':

> I say extravagances for what other name shall we give to such a
> chain of conjectures, based on opposites and resemblances so
> remote that the dreams of the sick-bed could not be more
> bizarre or disconnected.

His biographer P. N. Furbank, says 'conjecture' was for
Diderot, this man of the Enlightenment, 'a sort of wilful
going blindfold'. And so, back and forth, till 'truce' –
'some kind of whole made of shivering fragments ...
symmetry by means of infinite discords' (Virginia
Woolf).[7] The painter, the patient, faces 'something
known and at the same time unknown' which he can
take cognisance of and befriend.

Images that come up in therapy, whether they are
painted, modelled or arise in dreams, fantasy or
memory, will become part of the therapeutic process,
enriching and inspiring it and will themselves be enriched within it. In analysis, it would give me much

pleasure when referring to images became a kind of code, when each image with its associations – the mountain, the journey, the face – helped to map out the psychic territory created between analyst and analysand. Images as passwords to gain speedy entry into different psychic domains.

On pages 44–5 and 104–7 I have written about the formative and curative power of images. Only recently have I found this to be an age-old wisdom, namely when I read in the Book of Numbers in the Old Testament that the brass image of a fiery serpent had the power to cure the people who had been plagued by the fiery serpents God had sent amongst them for speaking against him. Moses prayed for the people.

> And the Lord said unto Moses, Make thee a fiery serpent and set it upon a pole: and it shall come to pass, that every one that is bitten, when he looketh upon it, shall live. And Moses made a serpent of brass, and if a serpent had bitten any man, when he beheld the serpent of brass, he lived.

In not a dissimilar way the analytical art psychotherapist Joy Schaverien[8] views pictures produced by patients as 'scapegoats' within the therapeutic relationship. It is in the pictures that serpents come forth; the pictures bear the burden of feelings and fantasies, possibly aggressive, destructive ones in regard to the therapist, and it is in the form of images that unacceptable things can be looked at with some safety and 'survived'. Here she is referring to D. W. Winnicott's chapter in *Playing and Reality* entitled 'The Use of an Object', a text that is illuminating about the creative process itself.

Winnicott came to the concept of Object Usage late in life, but the ground may have been laid for it when during the war he looked after hostels set up for evacuated children. He found that after an initial period of responding to the warden and staff as if they were ideal father and mother figures, the child had to test them. And this had to be allowed for.

> [The child] wants to know what damage he can do, and how much he can do with impunity. Then if he finds that he can be physically managed, that is, that the place and the people have nothing to fear from him physically, he starts to test by subtlety [and gets up to all sorts of tricks] ... And finally, if the hostel withstands these tests the child enters the third phase, settles down with a sigh of relief, and joins in the life of the group as an ordinary member.[9]

This was in 1940 or so. Winnicott first formulated *The Use of an Object* in 1968, long after he had written about *Transitional Objects* which was in 1951.

I think in the infant's life Transitional Objects and what he calls the Use of an Object overlap in time. Both appear after an initial period of perhaps up to six months during which the infant exists in a reciprocal primary relationship merged in with the mother in an unintegrated state. But even at the unintegrated stage aggression is part of the infant's appetite, greed, part of the primitive expression of love, of motility, of what Winnicott calls purpose without concern. At this point the mother is a subjective object and by her compliance the baby is allowed the illusion that she is under his omnipotent control.

As the mother's adaptation becomes less complete and the baby goes at her with all its aggression and love again and again as if to destroy her, eat her up, and if yet the mother is robust enough, puts up with and survives these attacks, then the baby gradually begins to understand that she is beyond his power and therefore fully real. By reflection the infant is also real, and some degree of independence is achieved. It is a definite developmental stage. A world of shared reality is created which the baby, the child, can use and use too in the sense of enjoy.

If things go wrong and the mother is angered by her baby's attacks and retaliates or withdraws in some way, then the baby may not dare to try again and will remain locked in a subjective world. In later life such a person in analysis may need to attack and test the analyst to find a way out.

It is the 'primitive force', as Winnicott called it, of these early days which the artist is able to be in touch with. His concept of Object Usage may be a way of understanding the destructiveness that is certainly part of making any work. Just think of the beautiful white of the canvas attacked by the first brush stroke and then by the ruthless reworkings throughout, because, in fact, without such drastic measures nothing would move forward. This destructiveness is survived – mostly – by the picture. And by reflection the painter's being is affirmed.

I have seen two states of a Rembrandt etching of the Crucifixion, an earlier and a final one. In the earlier one the three crosses, roughly in the centre, as I remember, were surrounded on all sides by onlookers. In the final state Rembrandt had blacked out all that crowd in order to highlight the Crucifixion. What is that moment, I ask myself, when taking up his tool and seeming to risk all, he eradicates half the plate with hundreds of strokes in order to get closer to his main concern? It is now a very dark etching with a luminous centre.

Michelangelo in the last week of his life hacked away

– and this for a second time – great parts of his Milan
Pietà, in order to reform it according to some inner
need. He wanted to lift and draw back the head of
Christ and to do that he had to hack into the bosom of
the Mary of the first state – there was no other marble
left. The regular, slanting grooves of his chisel tell of the
vehemence and rhythm of his work. These figures, with
part of the wreckage of earlier versions, stand erect and
lofty. Three days later Michelangelo died.

All this to some degree underlies the creative making
that goes on in art therapy. We as art therapists are
confronted with the picture. It 'deals' with us in the
sense in which the author of *The Cloud of the Unknowing*
uses the word (p. 66). We are impressed, struck; it
rummages in the chest; it is as if our body receives it.
First and foremost it is a physiological communication.
The art therapist who has accompanied this process is
not only a witness, she may also be a touchstone.

Remembering earlier paintings by the same painter
we may discern that the tone has lightened, the colours
previously held apart have been allowed to merge, that
merging colours have been separated – even if only
here and there; lines have appeared; we see smooth
surfaces roughened, the rough refined. Where previously every inch of the page was filled and overfilled
we see spaces left; where marks and patches stood in
isolation they have drawn together. The placing of a
single red dot in a blue landscape-wash has poignancy,
it brought tears to the eyes of the patient. What does it
touch? The very making of a painting here is an enactment.

R. M. Simon, an art therapist of fifty years standing, looks at the whole Gestalt and style of a painting and finds that the style itself is symbolic of a patient's psychological attitude. From her observation of the developmental stages of children's art she has come to recognise four distinct styles ranging between the Archaic and the Traditional, the Massive and the Linear and thinks of these as placed in the four quarters of a circle with intermediate 'transitional' styles between them. Utterly attuned to recognising these styles and shifts from one style to another in her patients' work she finds her circle of styles a useful frame for checking her intuitive responses, much in the way that Susan Bach (p. 84) used her quadrant theory. 'It helps me', says Rita Simon. 'Not everybody finds it helpful', she adds quite humbly.

This seems a good attitude to theory. As we know, theories and systems may be fine constructions but we cannot live in them; at best we may camp close by. I feel affection for Freud when after setting out – and this for the second time – to elucidate the complications of psychic functioning, he postulates a mental apparatus with id, ego and superego. He presented his idea in diagrammatic form but admitted that he could not do justice to the psychic processes by linear contours: merging fields of colour would give a better idea. In fact we must allow the separated parts to merge again – it was only an attempt at clarification. It was a fiction.[10]

Rita Simon is very wary of 'unmasking' images by a direct reference to symbolic meaning. 'One reason for this is that I am always aware that a symbolic image however obvious its meaning might appear to be, actually carries a multiplicity of meanings.'[11] One may be able to sense such submerged meanings, she says, but

find it difficult to communicate these in words; even
were it possible, the danger is that words fix meaning
and might thus rob the symbol of its momentum. 'A
wandering root of multiple meaning animates poetic
speech' writes the poet Osip Mandelstam.[12] I like to
think of a wandering root of multiple meaning animat-
ing the art therapy process.

There is much work going on nowadays that allows
for this. I shall only have room to describe one such.

The art therapist Patricia O'Sullivan was entrusted to
set up the Thornbury Day Centre in Leyton, where the
old Langthorne Hospital has been pulled down. The
majority of people who attend have come through
Claybury, the old psychiatric hospital, and now live in
the community. They have elected to call themselves
'members'. Patricia O'Sullivan feels that for far too long
they have been seen as patients needing treatment, and
is inclined to regard their breakdowns, crisis situations,
or very real sorrows as an opportunity, a starting point
towards recovery. 'Even when ill, there is so much
health and good in a person, it is this part I like to
address.' Drama therapy, occupational therapy, garden-
ing, relaxation, massage, counselling are on offer. Art
therapy is practised in an open studio, in closed groups
and in one-to-one sessions. In days where art therapy is
becoming more and more highly defined and protected,
she moves in the opposite direction. She told me on my
visit to this very lively and expanding Centre that she
made a point of always inviting someone from another
discipline – a counsellor, a nurse, a therapist in another
field – to attend the closed group she runs. She welcomes and values the different perspective such a person will bring. Conversely she feels that there is so
much to be learned from the group that it made for an

enrichment in her colleagues' work, thus spreading the gain throughout the Centre.

Multiple perspectives are also a feature in Shaun McNiff's method.[13] He, an American art therapist and professor of Expressive Therapies, much infuenced by the imaginal work of James Hillman, nurtures images like a gardener, somewhat of a hothouse gardener maybe, attending to propagation, seedlings, offshoots and flowering. The dialoguing with the image that he practises with his groups is aligned with the art process and will foster the inherent procreative power of the image.

> Rather than being defined by boundaries, reality seems closer to a continuous process of interaction. Physics and the arts support this interactive vision, which suggests that our therapeutic purpose is to further the sensitivity and imaginative scope of the interplay.

For Lyddiatt, it may be remembered, the forging of a relationship with the archaic layer of the mind – 'the living creative matrix' – as revealed in the spontaneous paintings and models made in her department, was paramount. I remember her, rather wise, her blue eyes wide open, wondering at the manifold disclosures of a painting. I seem to have spiralled back to my starting point.

Now, however, that art therapy has incorporated psychotherapy – and both the training and the literature reflect this – the picture a patient will paint is regarded as a focus for transference and countertransference and is studied in that light; that is, a

constant lookout is kept as to what attitudes acquired in
early development are played out (particularly in the
picures) and what reaction these attitudes call up in the
therapist. A sensitive practitioner will readily admit
that this scrutiny does not exhaust the image, that there
is in fact much more to it, and yet, by defining what is
to be looked for, it appears to me that what is not being
looked for is in danger of being atrophied. It has no
witness and willy-nilly there is an impoverishment, the
pictures tend to exemplify a linear process. And how
could this not be? (If my friend makes a fuss of certain
clothes I put on, I will leave the ones that get no notice
in the cupboard. Perhaps Lyddiatt had seen the logic of
this when she asked for there to be 'no comment'.) And
perhaps it doesn't matter since the centre of the work
has shifted into the therapeutic alliance.

Victor Turner,[14] the anthropologist and author of *The
Forest of Symbols* distinguishes these two modes: liminality (from *limen*, Latin for threshold) and structure,
which follow each other and of necessity form a dialec-
tic in society. Thus, for instance, St Francis, humble and
poor, living his faith, was superseded by the well-
endowed Franciscan Order, where structure both
material and abstract enshrined him.

For St Francis, so I read, a sequence of thought
consisted in leaping from one picture to the next; ab-
stractions were alien to him; the multivocality of his
symbols made him a poor legislator; he had no gift for
generalisations, classifications. He remained informed
by that fundamental working of the mind, the soul,
where images lead – toss you about in the case of
Diderot. The French philosopher and phenomenologist
Gaston Bachelard[15] observed that studying something
was often no more than being receptive to a kind of

day-dreaming. Images are the kernal of our thought. Only later do we begin to see patterns and form, and this is how art therapy theory has evolved; we begin to discern what may be a law. We try to formulate it and hold it in words.

Under liminality Turner lists totality, simplicity, silence, humility and foolishness and confronts it with the properties of status and structure: partiality, complexity, speech, just pride of position, sagacity.

Our work is on thresholds and its positive generative force is anti-structural. How to treasure this, now that that dangerous word technique holds sway, now we have methods and structure? First of all, perhaps, by admitting that we are 'horseless' each time we are face to face with the person we work with. Possibly by painting, making sculpture ... where we would immediately be steeped in an imagist way of thinking. But there is a danger here. Art will lay its claim. And since this book has been somewhat of a travel journal, I must report that the questions I raised at the outset – is she an artist? Is she a therapist? Or both? – still tease me. I have become a potter. I build pots.

Martina Thomson
May 1997

❧ SOURCES in the Postscript

1. Susan Howe, *My Emily Dickinson* (Berkeley, California: North Atlantic Books 1985) p. 23.
2. Samuel Beckett in *Avigdor Arikha* by various authors (London: Hermann publishers in arts and science 1985) p. 10.
3. Robert Hughes, *Frank Auerbach* (London: Thames and Hudson Ltd 1990) p. 196.
4. *Degas by himself*, ed. Richard Kendall (London: Macdonald Orbis 1987) p. 319.
5. Constantin Stanislavsky, *An Actor Prepares* (London: Geoffrey Bles 1945) pp. 293–4, 282.
6. P. N. Furbank, *Diderot, A Critical Biography* (London: Secker and Warburg 1992) pp. 111, 113.
7. From a notebook kept by Virginia Woolf on her journey to Italy in 1908. Quoted by Quentin Bell in *Virginia Woolf. A Biography* (St Albans, Herts: Triad/Paladin 1976) Vol. 1, p. 138.
8. Joy Schaverien, *The Revealing Image* (London: Routledge 1992) pp. 30–7.
9. Adam Phillips, *Winnicott* (Fontana Press, HarperCollins Publishers 1988) p. 68.
10. Sigmund Freud, *New Introductory Lectures on Psycho-Analysis. The Standard Edition of the Complete Psychological Works* (London: The Hogarth Press and the Institute of Psycho-Analysis 1964) p. 79.
11. R. M. Simon, *Symbolic Images in Art As Therapy* (London: Routledge 1997) p. 68.
12. *Osip Mandelstam: Selected Essays*, translated by Sidney Monas (Austin and London: University of Texas Press 1977) p. 81.
13. Shaun McNiff, *Ethics and the Autonomy of Images, The Arts in Psychotherapy* (USA: Pergamon Press plc 1991) Vol. 18, p. 280.
14. Victor W. Turner, *The Ritual Process: Structure and Anti-Structure* (London: Routledge and Kegan Paul 1969) pp. 141–2, 106–7.
15. Gaston Bachelard, *The Poetics of Space*, translated from the French by Maria Jolas (Boston, Massachusetts: Beacon Press 1969) p. XXXIV.

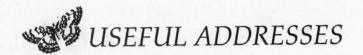 *USEFUL ADDRESSES*

The British Association of Art Therapists
11a Richmond Road, Brighton, Sussex BN2 3RL

Art Therapy Training Courses

Edinburgh University Settlement
School of Art Therapy
7/9 South College Street, Edinburgh EH8 9AA

University of Hertfordshire
School of Art and Design
Manor Road, Hatfield, Herts AL10 9TL

Goldsmiths College, University of London,
Art Psychotherapy Unit
23 St James', New Cross, London SE14 6AD

University of Sheffield
Centre for Psychotherapeutic Studies
16 Claremount Crescent, Sheffield S10 2TA

City of Bath College
Dept of Visual Arts
Avon Street, Bath, Avon BA1 1UP

 *INDEX*

'Active Imagination' 19, 21, 105
Adamson, Edward
   healing art 22, 49
   methods 40, 52–3
   spontaneous art 23, 25
Art Brut (Raw Art) 24, 38
art teaching 5–11
art therapists
   artists as therapists 2, 3–4, 131
   as facilitators 15, 22, 25, 63
   minimum intervention by 33, 48, 102, 108
   as observer/partici-pants 17, 30, 40–2, 58–9
art therapy
   development of 54–9, 118–19, 131
   directive/non-directive 50–1, 52–3, 57–8
   and psychotherapy 2, 21, 129–30
   theory 17–18, 127
   training 49–56, 129

Bach, Susan 83–4, 96–7, 99, 127
Beckett, Samuel 43, 120
Binswanger, Ludwig 42, 69–70
'blank page' 60–1
Bomberg, D. 7–10, 95
boundaries *see* lines
British Association of Art Therapists 13, 22, 49, 54, 118

chromotherapy 83
Collins, Cecil 81
colour 7, 56, 74–85
   black 69, 79–81
   red 29–30, 69, 75, 84–5
creative process 19–20, 47, 62–7, 99–100, 126, *see also* Active Imagination

destructiveness in creative process 63, 123–6
developmental stages in children's art 127
diagnosis *see* paintings/ models

Dubuffet, Jean 24, 38

Ehrenzweig, Anton 63–4,
72, 99–101
'dedifferentiation' 63–4,
90, 91
evaluation *see* paintings/
models
evidence-based medicine
(EBM) 119

formlessness 46–8, 102
Freud, Sigmund 26, 46, 127
archaic heritage 18, 98

Guggenbühl-Craig, Adolf
10, 11

Hegel, Georg Wilhelm
Friedrich 67, 101–2
Hill, Adrian 3, 25
Hillman, James 107, 129

image
power of 44–5, 104–8,
130–1
in therapeutic process
122–3, 129–30
interpretation, *see*
paintings/models
Itten, Prof. Johannes 6–7,
81

Janet, Pierre 42–3

Jung, C. G.
Active Imagination
19–21, 37, 105
patients' paintings 101
unconscious 17

Kandinsky, Wassily 77–8,
85
Khan, Masud 46–7, 102
Klee, Paul 65–6, 92
Kohut, Heinz 46

Lacan, Jacques 44–5
Layard, John 66
lines 89–93
Lydiatt, E. M. 13, 14–21,
37, 56, 60, 100–1
art therapy methods
15, 25, 33, 47
collective unconscious
18–19, 98, 129
departments 15–17, 22

Maclagan, David 34
McNeilly, Gerry 58, 59,
60
McNiff, Shaun 129
materials *see* media
Matisse, Henri 87, 92, 93,
97, 102
Meares, Ainslie 29–30, 43
media/materials 49–50,
62, 68–73, 86
mental illness 64

Merleau-Ponty, Maurice 77, 82, 83
mess 71–3, 86
Milner, Marion 30–2, 47–8, 90–1
'mirror phase' 44–5
modelling 86–8, 94–5, 105
music 74–5, 77–8, 82, 85

Navratil, Dr Leo 26–8, 33, 38
Nicholson, Ben 6

object usage 123–5
O'Sullivan, Patricia 128
outlines and boundaries 90, 92, *see also* lines
Outsiders Exhibition (Hayward Gallery) 24, 31, 33

paintings/models
    collections of 23–4, 36
    diagnosis from 83–4, 96–7
    evaluation of 37–9, 40, 87, 103
    interpretation of 22, 99–103, 127–8, 129–30
    owning up to 99–101, 104
    recognition 93–8, 104–6, 126
Pankov, Gisela 87–8

Picasso, Pablo 34–5, 91, 100
    on psychotic art 38–9, 92
Prinzhorn, Hans 23–4, 26, 33, 38
psychotherapy 2, 21, 33–4, 129
psychotic art 38–9, 91–2
    see also schizophrenia

recognition *see* paintings/ models
Riley, Bridget 95, 107
Rosen, John 27

Schaverien, Joy 123
schizophrenia 26–8, 38, 39, 91–2
Schopenhauer, Arthur 67, 95
Sechehaye, Marguerite 27–8, 43, 57
secret art 23–5, 31–2, 34–6, 100
'selfing' 48
Simon, R. M. 127–8
solitude, need for 35–6
sound and colour 77–8
'spirit in the mass' 8, 95
spontaneous art 3, 23, 29, 31, 33
    doodles 31, 93
Steiner, Rudolf 56, 75

symbolism 28–30, 32, 43,
98, 127–8
in colour 69, 76, 78–81,
84–5

Thomson, Martina
background 4
introduction to art
therapy 12–13, 15–17
training 49–53
transference 2, 103, 118,
129–30
Turner, Victor 130, 131

unconscious 17–18, 36,
63–4 65, 102, 121–2
collective unconscious
18–19, 97–8, 129

Westman, Heinz 30, 41
Winnicott, D. W. 58, 102,
103
infant development
43–5, 46, 123–5
Wolfli, Adolf 34–6

Index by Sue Carlton